3D PRINTING WITH AUTODESK® 123D®

CREATE AND PRINT 3D OBJECTS WITH 123D, AUTOCAD, AND INVENTOR

John Biehler
Bill Fane

que®

800 East 96th Street,
Indianapolis, Indiana 46240 USA

3D PRINTING WITH AUTODESK® 123D®

ISBN-13: 978-0-7897-5328-1
ISBN-10: 0-7897-5328-6

Library of Congress Control Number: 2014938539

Printed in the United States of America

First printing May 2014

Editor-in-Chief
Greg Wiegand

Executive Editor
Rick Kughen

Development Editor
Sondra Scott

Managing Editor
Sandra Schroeder

Project Editor
Seth Kerney

Copy Editor
Megan Wade-Taxter

Indexer
Lisa Stumpf

Proofreader
Kathy Ruiz

Technical Editor
Ralph Grabowski

Publishing Coordinator
Kristen Watterson

Cover and Interior Designer
Mark Shirar

Compositor
Mary Sudul

Contents at a Glance

Table of Contents

About the Authors

John Biehler has been writing online about technology since 1999. An avid photographer and generally curious geek, he discovered 3D printing a number of years ago and built his first 3D printer shortly thereafter. Since then, he has been actively sharing his knowledge about the technology with thousands of people at various events and conferences in Western Canada and the Pacific Northwest, on television and radio, as well as online through his website. He cofounded a Vancouver-area group of 3D printer builders and enthusiasts that has grown exponentially since it started and as the technology heads toward the mainstream.

Bill Fane was a product engineer and then product engineering manager for Weiser Lock in Vancouver, British Columbia, for 27 years and holds 12 U.S. patents. He has been using AutoCAD for design work since Version 2.17g (1986) and Inventor since version 1.0 beta (1996). He is a retired Professional Engineer and an Autodesk Authorized Training Centre (ATC) certified instructor. He began teaching mechanical design in 1996 at the British Columbia Institute of Technology (BCIT) in Vancouver, including such courses as AutoCAD, Mechanical Desktop, Inventor, SolidWorks, machine design, term projects, manufacturing processes, and design procedures. He retired from this position in 2008. He has lectured on a wide range of AutoCAD and Inventor subjects at Autodesk University since 1995 and at Destination Desktop since 2003. He was the AUGI CAD Camp National Team instructor for the manufacturing track. He has written more than 220 "The Learning Curve" AutoCAD tutorial columns for *CADalyst* magazine since 1986. He is the current author of the book *AutoCAD for Dummies*. He also writes software product reviews for *CADalyst*, *Design Product News*, and *Machine Design*. He is an active member of the Vancouver AutoCAD Users Society, "the world's oldest and most dangerous." In his spare time he skis, water skis, windsurfs, scuba dives, sails a Hobie Cat, rides an off-road motorcycle, drives his '37 Rolls Royce limousine or his wife's '89 Bentley Turbo R, travels extensively with his wife, and plays with his grandchildren.

Dedication

John Biehler: I want to dedicate this book to Kelli Smith, who watched patiently as I built my first 3D printer on my dining room table and has supported my efforts in 3D printing ever since my first printed object came off the printer.

Bill Fane: To my wife Bev, who still manages to tolerate me after being married for 48 years.

Acknowledgments

John Biehler: I want to acknowledge and thank friends and members of 3D604.org, the Vancouver Hackspace, and Metrix Create:Space in Seattle. Without their help, friendship, and willingness to share their knowledge, my participation in this book would not have been possible.

We Want to Hear from You!

As the reader of this book, *you* are our most important critic and commentator. We value your opinion and want to know what we're doing right, what we could do better, what areas you'd like to see us publish in, and any other words of wisdom you're willing to pass our way.

We welcome your comments. You can email or write to let us know what you did or didn't like about this book—as well as what we can do to make our books better.

Please note that we cannot help you with technical problems related to the topic of this book.

When you write, please be sure to include this book's title and author as well as your name and email address. We will carefully review your comments and share them with the author and editors who worked on the book.

Email: feedback@quepublishing.com

Mail: Que Publishing
ATTN: Reader Feedback
800 East 96th Street
Indianapolis, IN 46240 USA

Reader Services

Visit our website and register this book at quepublishing.com/register for convenient access to any updates, downloads, or errata that might be available for this book.

The Rise of 3D Printing

The Industrial Revolution of the late 1700s through to the mid-1800s brought about a radical change in the social and economic fabric of society. It was so radical a change that analysts, even a few decades earlier, didn't see it coming.

Before then, agriculture occupied well over 90% of the population and individuals fabricated nearly all human-made objects one at a time. In fact, in many cases it was the end user of a product who handled the fabrication.

The Industrial Revolution began the shift away from hand-made products to machine-made ones, the use of coal instead of wood, new chemical processes, and new processes for refining iron and making steel.

This continued through into the early twenty-first century. As you read this book, look around you. How many objects, if any, can you see that were hand-crafted? How many, if any, did you make?

This revolution in manufacturing means that average people in the industrialized nations today live better than kings and emperors of less than a century and a half ago. If you are sitting on the toilet as you read this, note that the flush toilet is only 150 years old.

TRIVIA

Queen Victoria had one of the first flush toilets. Her plumber was Thomas (later "Sir Thomas") Crapper and hence the origin of the phrase "To take a crap."

In 1984, Charles Hull invented *stereolithography,* which is derived from the Greek words *stereo* (solid body), *litho* (stone), and *graphien* (to write). Stereolithography is really just a more scientific term for 3D printing. In a relatively short period of time, stereolithography is becoming as revolutionizing as the Industrial Revolution— and, no, contrary to popular opinion, Jay Leno didn't invent it on his TV series, *Jay Leno's Garage*, even though he showed it being used to reproduce a part for one of his many collector cars.

Stereolithography is just one term that has been used to describe 3D printing. It has also evolved through several other names in its short life including rapid prototyping, additive manufacturing, and rapid manufacturing. Personally, I prefer Captain Kirk and company, who called it the "matter replicator," but for the purposes of this book we'll go with 3D printing.

3D Printing Will Change the World

It's that déjà-vu thing all over again. Like other revolutions, 3D printing is going to revolutionize how you live, and sooner than you think. No more going to the store; just go online, buy a file, download it to your printer, and out comes the object you want. What if you can't find anything exactly to your liking? No problem, just design it yourself and print it. Personalized mementos for weddings, anniversaries, birthday parties, or corporate events? No problem. Need a part for the vintage car you're restoring? Take 20–30 digital photos, upload them to a free website and in less than a minute you get back a file of a 3D model that can be fed to your 3D printer.

→ *Look through Chapter 4, "Creating 3D Objects with Cameras and 123D Catch," to see some images of parts created with 3D printing.*

Even though we don't have the matter replicators that were used in the original *Star Trek* series, we have mobile "communicators" that we call cell phones. A telephone that lets you see the other person was science fiction just a few years ago, while today we regularly Skype (for free!) between continents.

The rate of change of technology since I was born in the early 1940s is amazing. The following inventions didn't exist or weren't in common usage:

Jet engines

Most plastics

Color TV

Transistors, integrated circuits, and computer chips

Computers

Satellites

Credit cards

Digital watches

Cable TV

Audio cassettes

VCRs, CDs, DVDs

Fax

Cell phones

Internet

Digital cameras

Many of you may not be old enough to remember a world that didn't have some of the items on this list. While you're at it, note how many of them have come and gone already.

If you still think the leap from 3D printing to *Star Trek*-style replicators is far-fetched, here are another few tidbits to help put the rate of change in perspective:

- Someone born 20 years before the Wright brothers flew may have lived to see men walk on the moon.
- Cell phones have more computing power than all of NASA had at its disposal when it put men on the moon.
- When the Xbox game console was introduced, it had more computing power than the biggest military super-computers of the day.

As I see it, the biggest problem facing science-fiction writers today is keeping ahead of technological changes.

With these changes in mind, it's easy to imagine how 3D printing will indeed change our world—sooner rather than later.

In the next several chapters we'll take you through some basic principles of 3D printing, followed by several home hobbyist chapters, then professional (that is, AutoCAD and Inventor) chapters, and finally hint at what might be coming in the not-so-distant future.

In the 1927 movie *The Jazz Singer*, which overnight revolutionized Hollywood, Al Jolsen began with "You ain't heard nothin' yet!" Well, hang on tight, because you ain't seen nothin' yet!

Basic Principles of 3D Printing

All 3D printing works on the same basic principle, but there are several variants and sub-variants. In this chapter we'll explain the basic principle and then go on to discuss the variants.

How 3D Printing Works

Before you can 3D print an object, you need a 3D computer model of the object. Many programs are available that can produce suitable models. For example, you can use engineering and design software such as AutoCAD or Inventor from Autodesk, SolidWorks from Dassault Systemes, or Solid Edge from Siemens PLM Systemes.

Okay, so you can't afford several thousand dollars' worth of engineering software. No problem; Autodesk has free software that you use online at www.123dapp.com. This even includes a killer app called 123D Catch that will produce a 3D model using 20–30 digital photos from an ordinary camera, tablet, cell phone, and so on.

NOTE

You can also download thousands of existing models from a number of websites. This is probably the easiest approach if you just want to see your new printer working right away. The flip side is that you don't even need a 3D printer. There are a number of online service bureaus. You upload your model—often as easily as clicking a "3D Print" button in your design app—and they ship you the part (for a nominal fee, of course). We cover both the engineering software and the 123D products in later chapters.

Layered Approach to 3D Printing

One feature of the 3D modelling software is that it can generate flat 2D drawings of 3D models, such as the cross-section views shown in Figure 2.1.

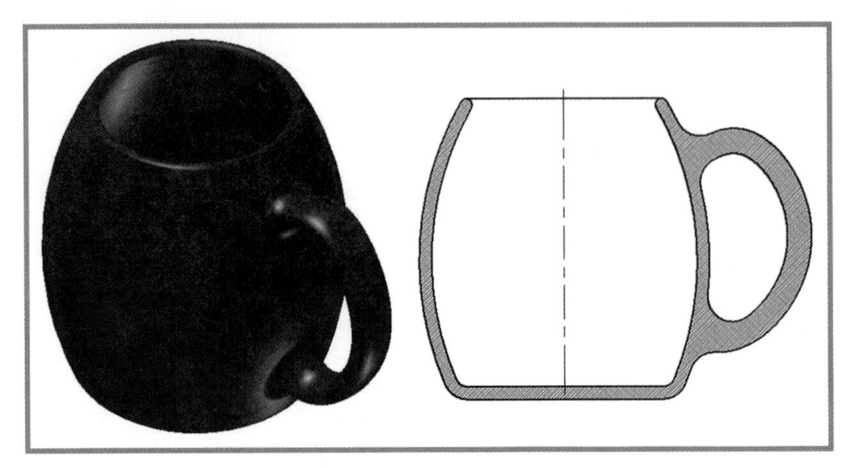

FIGURE 2.1 Cross-section made from a 3D model.

In a nutshell, that is the essence of 3D printing because all 3D printers work basically the same way. Their software takes in the 3D model from your design software, creates a cross-section view just a hair's width up from the bottom of the part, and then prints it as a paper-thin layer of solid material onto a flat platen. The print head lays down a very narrow stripe of material as it traverses back and forth, advancing sideways a tiny amount between each traverse.

NOTE

The only difference between all the brands and models lies in the type of material they use and how they print it. We'll have more on this later in this chapter.

Because the printing sequence for the mug shown in Figure 2.1 is more complex, let's begin by examining a simpler model of the letter *H*.

Figure 2.2 shows the first three stripes in blue while the yellow print head is working on the fourth stripe, shown in green. In the real world all stripes would normally be the same color and would not have dark edges. I added the edges and the green to make the stripes more obvious.

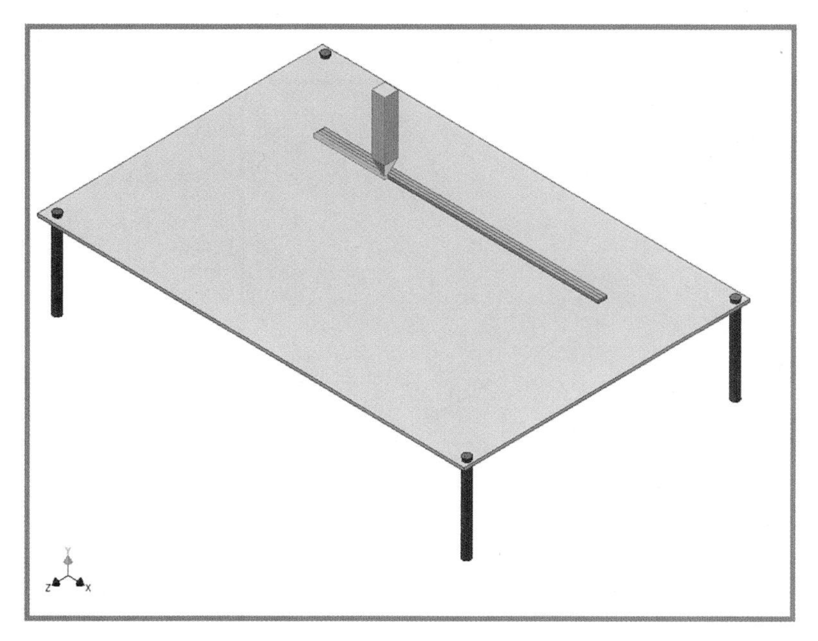

FIGURE 2.2 Working on the fourth stripe.

Figure 2.3 shows the half-finished first layer. The print head is working from right to left on the green stripe.

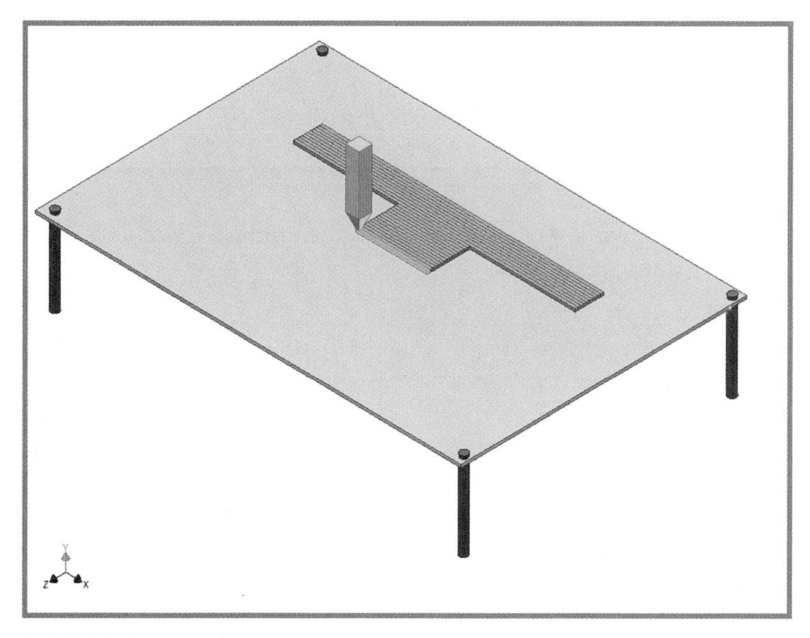

FIGURE 2.3 The first layer is half done.

Moving right along, Figure 2.4 shows the completed first layer.

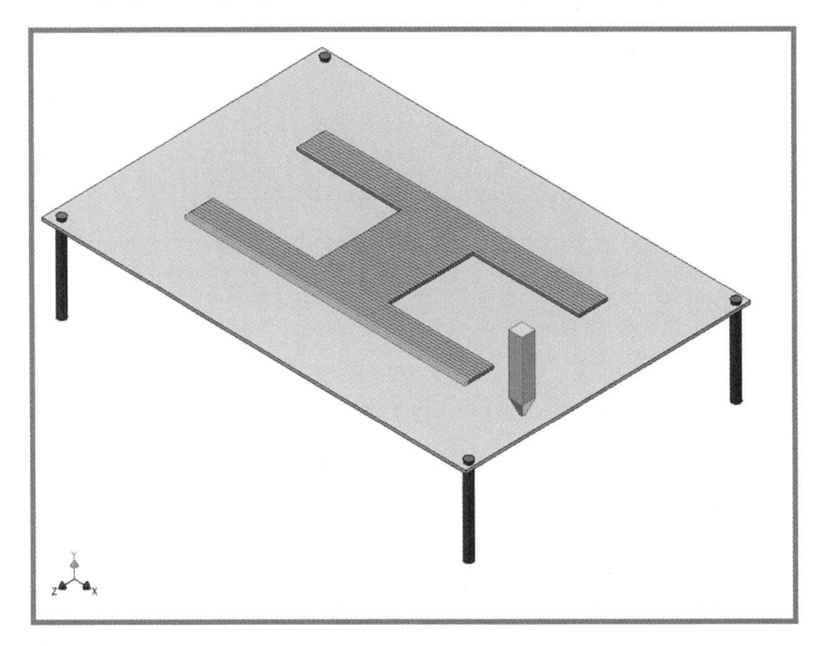

FIGURE 2.4 More stripes than a zebra! The bottom layer is complete.

The printer lowers the plate with the first layer on it by an amount equal to the thickness of the layer and moves the cutting plane in the 3D model up by the same amount. It then prints the new cross-section on top of the previous one, as shown in Figure 2.5.

TIP

Once again, for clarity, I've changed a color. The edges of the stripes remain dark, the current stripe remains green, and the current layer remains blue, but previous layers are light gray.

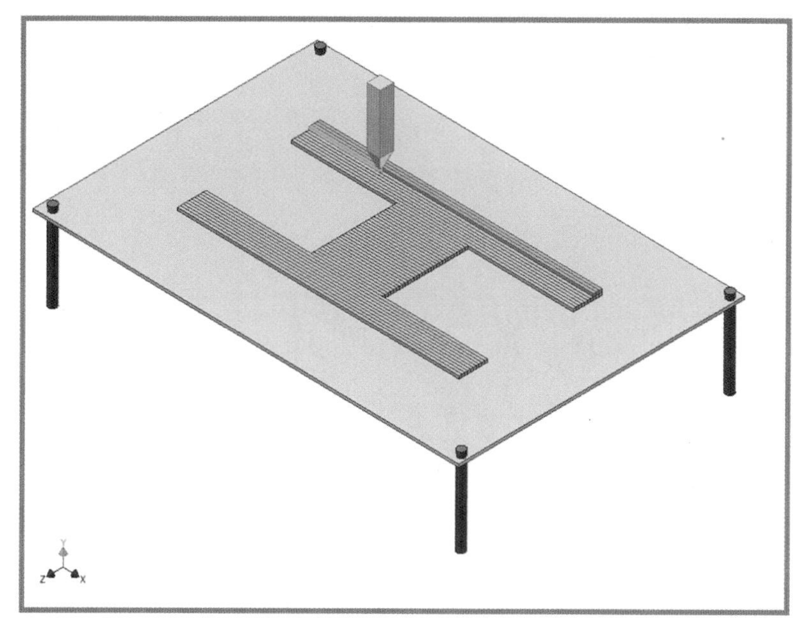

FIGURE 2.5 Starting the next layer.

The process continues. Figure 2.6 shows the half-finished second layer, and Figure 2.7 shows how the first two layers look when finished.

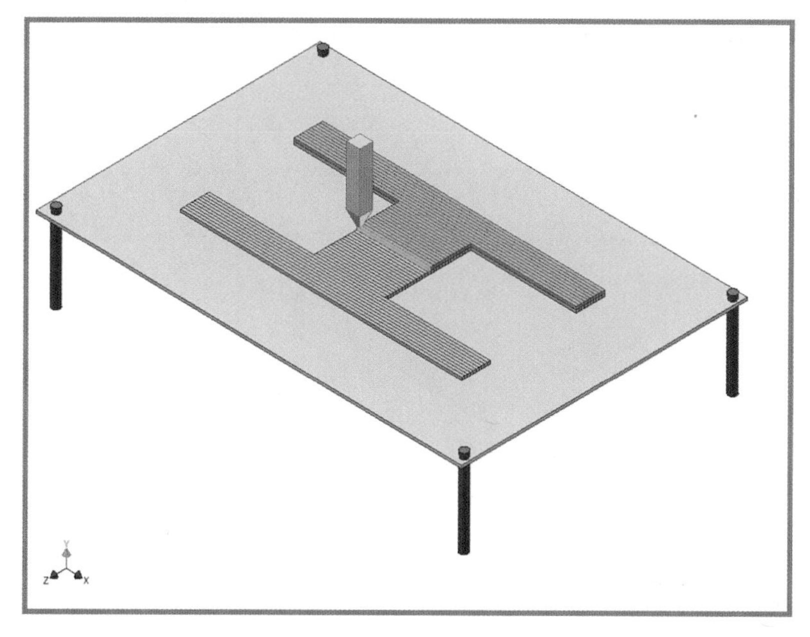

FIGURE 2.6 Working on the second layer.

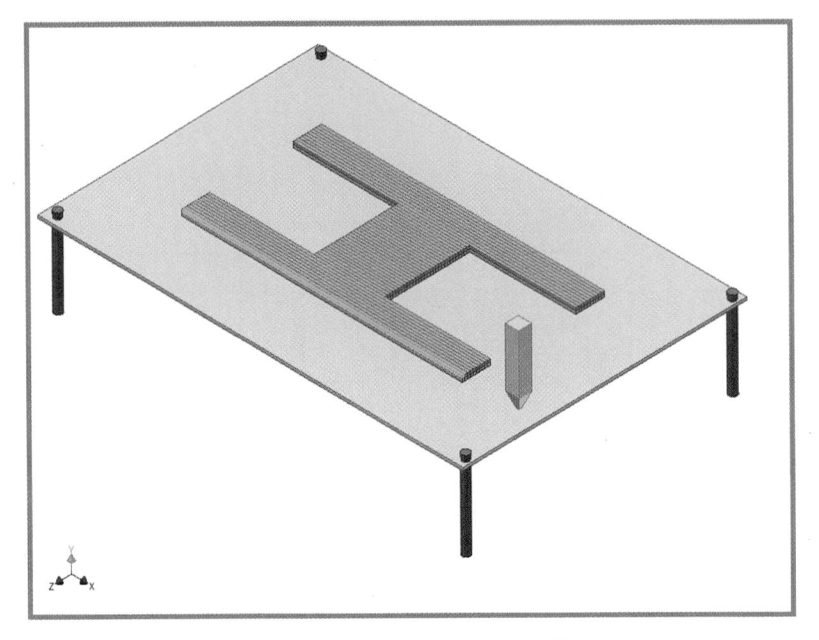

FIGURE 2.7 The first two layers are finished!

The process repeats, layer after layer, as shown in Figures 2.8–2.19, as the part seems to grow downward from the print head.

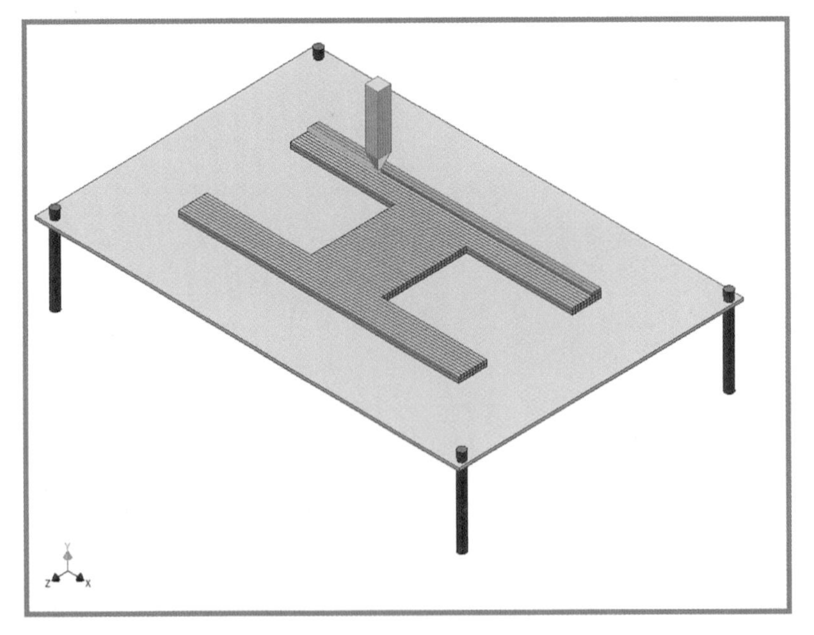

FIGURE 2.8 Starting the third layer.

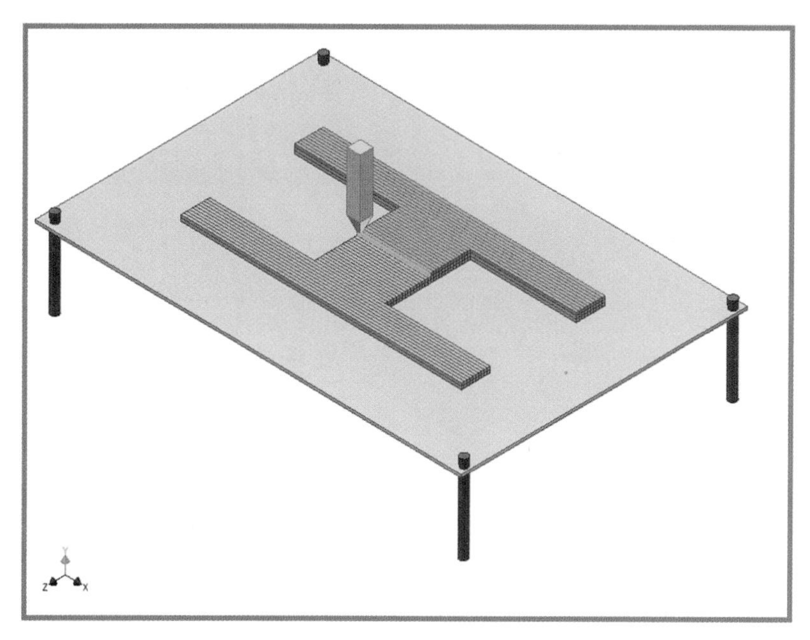

FIGURE 2.9 The third layer is half done.

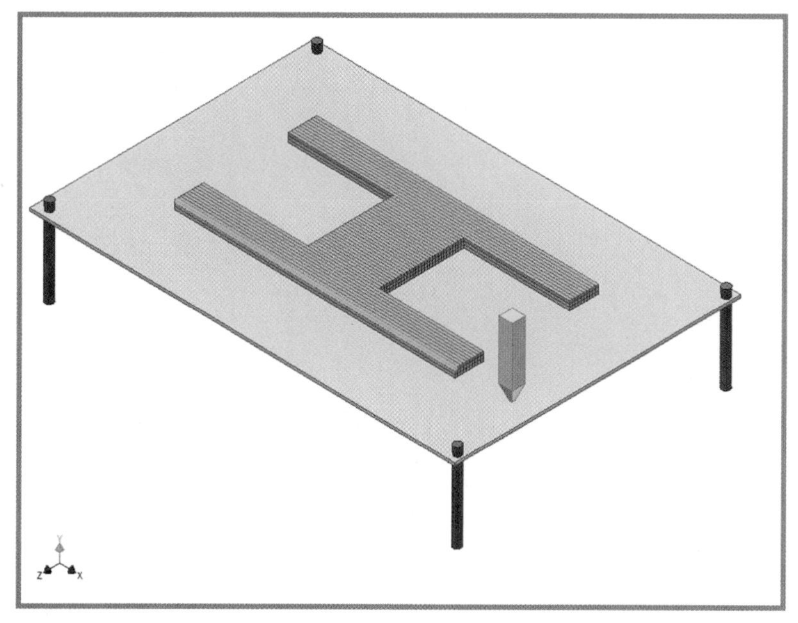

FIGURE 2.10 The third layer is finished.

I won't bore you with figures of every layer being created, so we'll jump ahead a bit to the tenth layer being started in Figure 2.11.

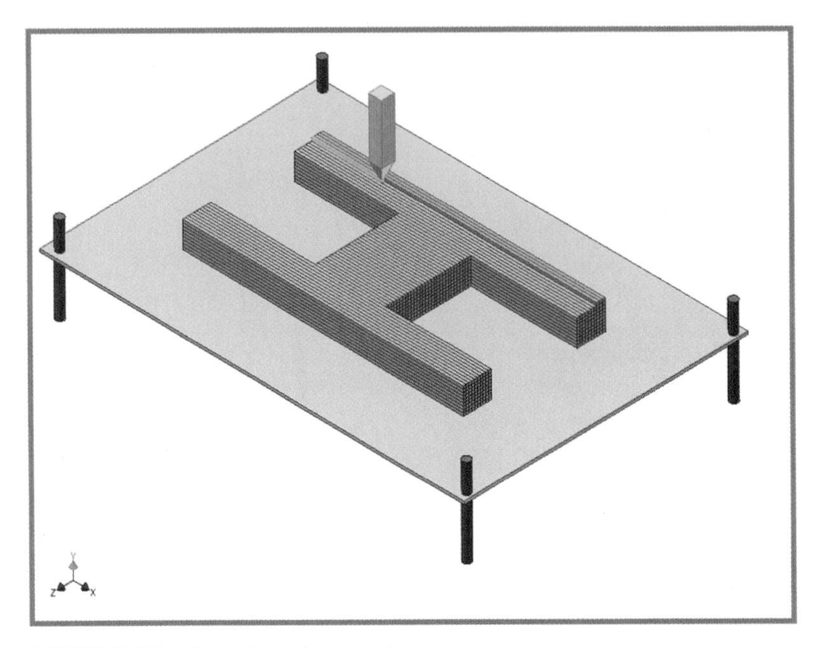

FIGURE 2.11 Starting the tenth layer.

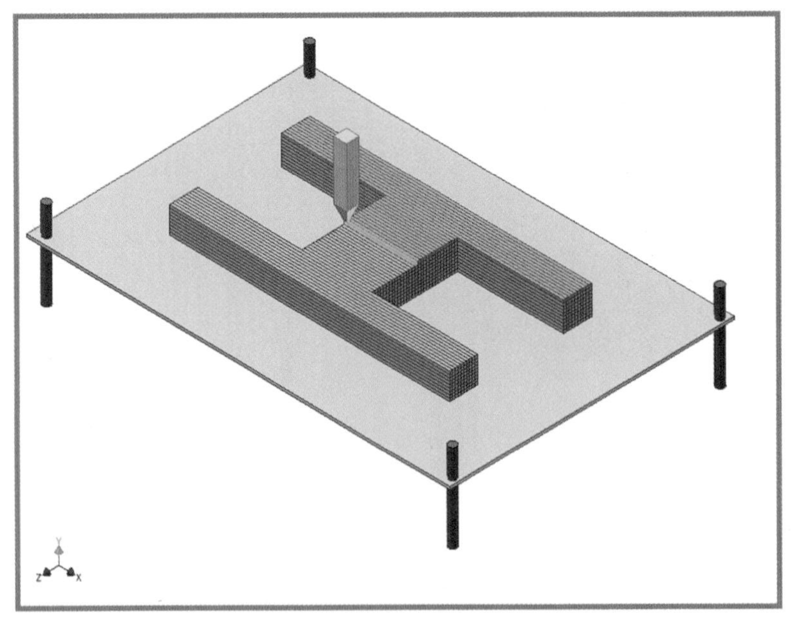

FIGURE 2.12 The tenth layer is half done.

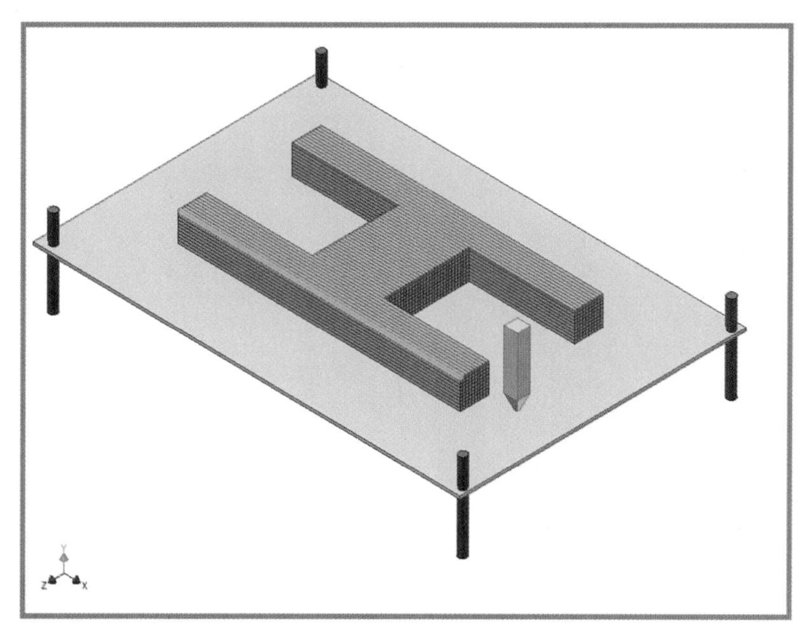

FIGURE 2.13 The tenth layer is finished.

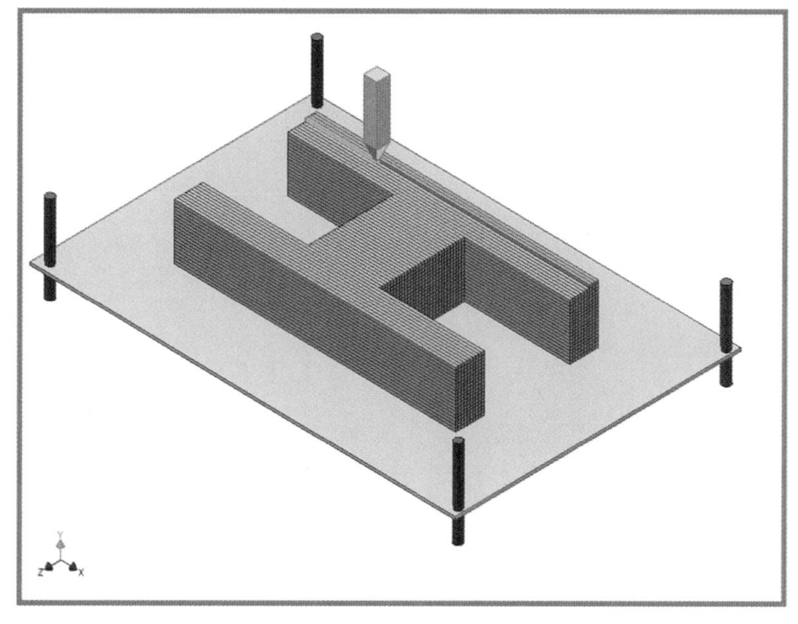

FIGURE 2.14 Moving on down, layer 20 starting.

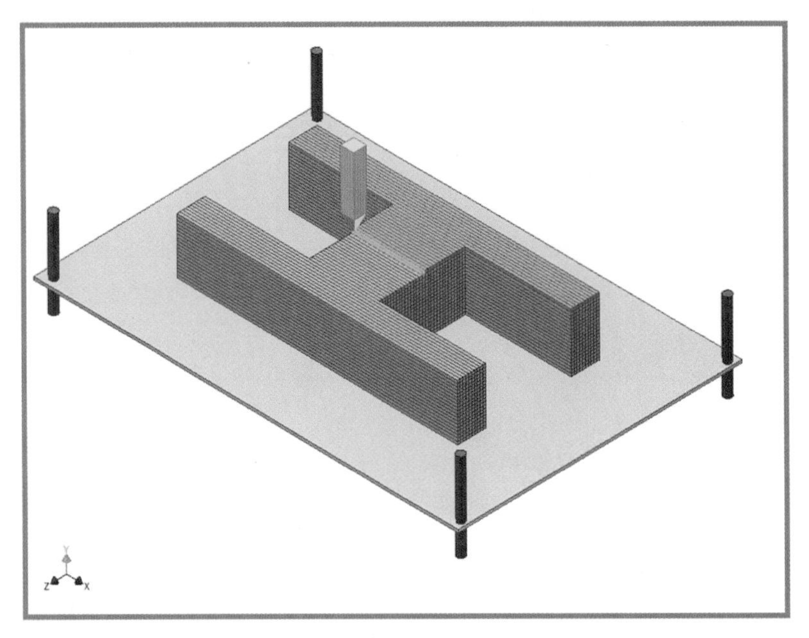

FIGURE 2.15 Layer 20 is half done.

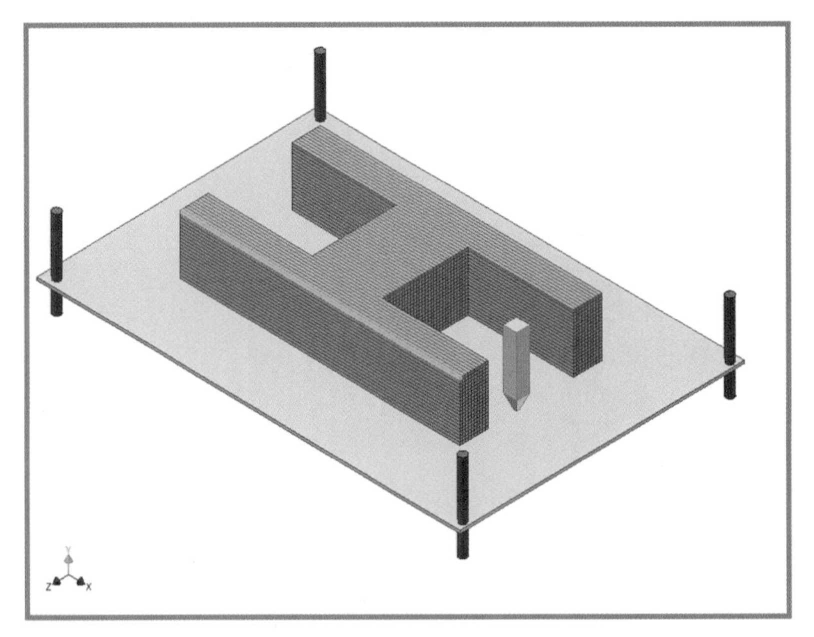

FIGURE 2.16 Layer 20 is finished.

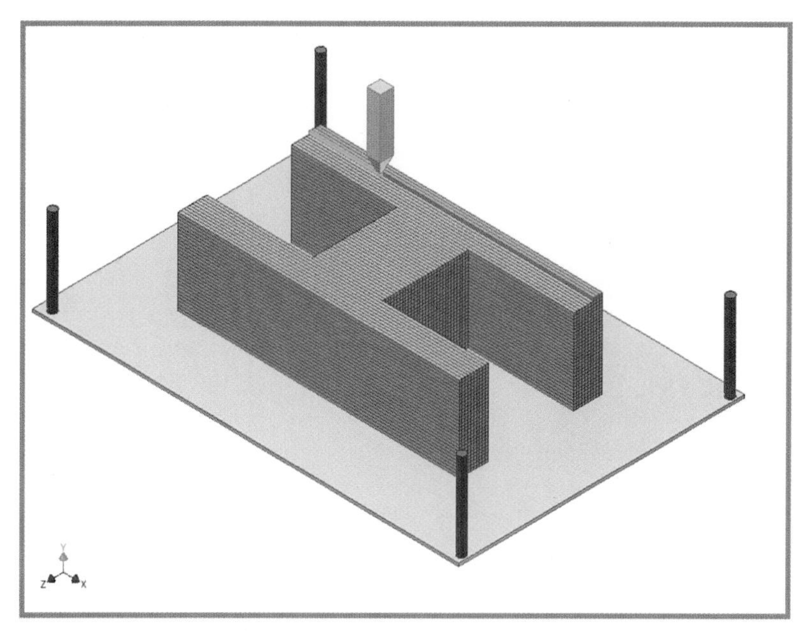

FIGURE 2.17 Layer 30 is starting.

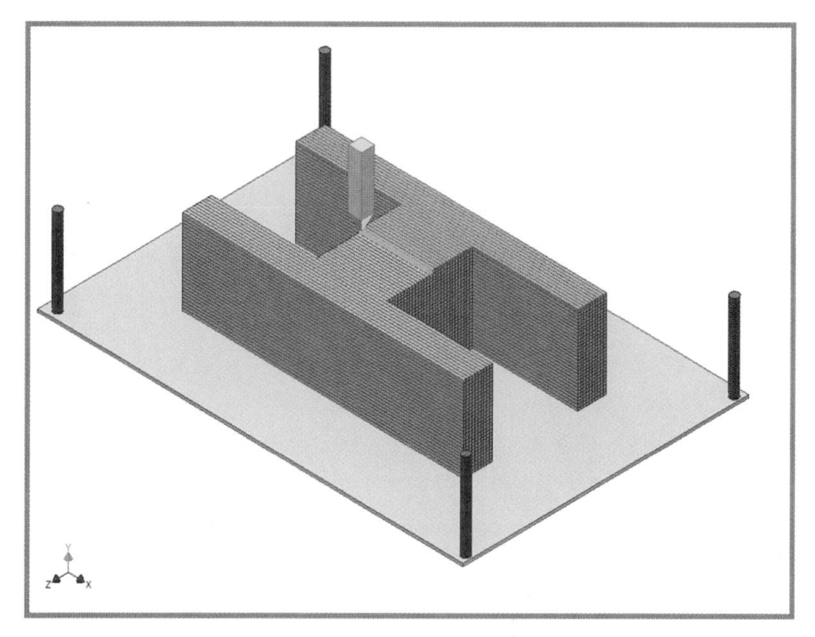

FIGURE 2.18 Layer 30 is half done.

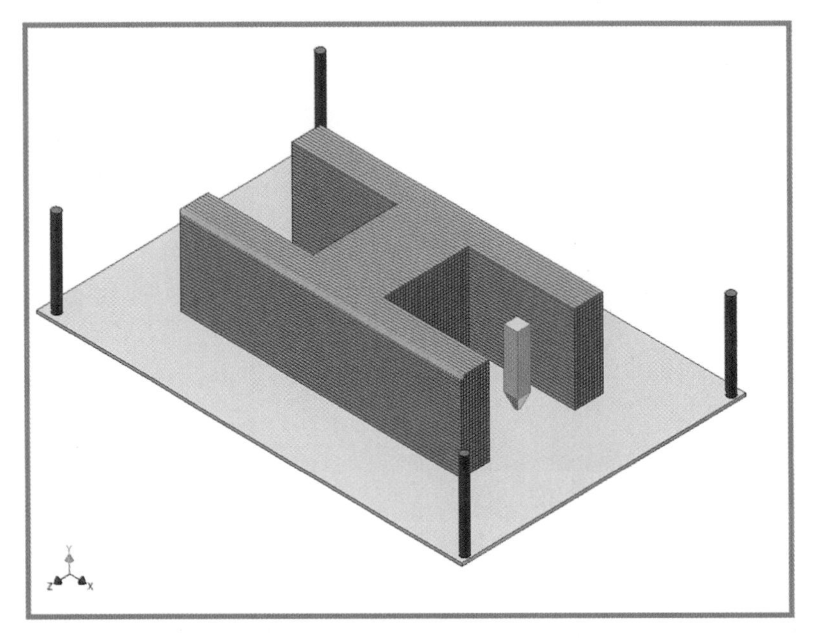

FIGURE 2.19 Thirty layers, and we're done!

If you want to get a better feel for how the 3D printing process operates, all you need to do is to tear the preceding pages out of the book, cut out the figures, and staple them together to make an animation flipbook. When you're finished, go buy another copy of the book because you just ruined the first one.

The next morning we can remove the finished part from the 3D printer, as shown in Figure 2.20.

FIGURE 2.20 Here's the finished part fresh out of the 3D printer.

Printing Time for Layers

"Did you say 'the next morning'?" That's right; the average 3D printer operates somewhat slower than your average laser printer, and they are often left running overnight. A significant point here is that printing time is not usually a function of the complexity of the part or of its width and length (within the machine's capabilities).

Instead, printing time depends mostly on the thickness of the part. Your car speedometer shows miles and/or kilometres per hour, but 3D printer speeds are typically in the range of an inch per hour in the vertical direction. Realigning the part before printing so that its shortest dimension is in the vertical direction can often significantly reduce the printing time.

TIP

For simplicity and ease of creating the figures, I made each layer identical. However, in the real world most 3D printers lay down succeeding layers in alternating alignments to minimize the possibility of "grain" effects in the final part.

Figure 2.21 shows this process in the first two layers of our previous example. The third layer would then be applied in the same direction as the first one, and this would repeat all the way through the part. If this was not done, then the part could easily be much weaker in one direction than another, much like the grain in a piece of wood.

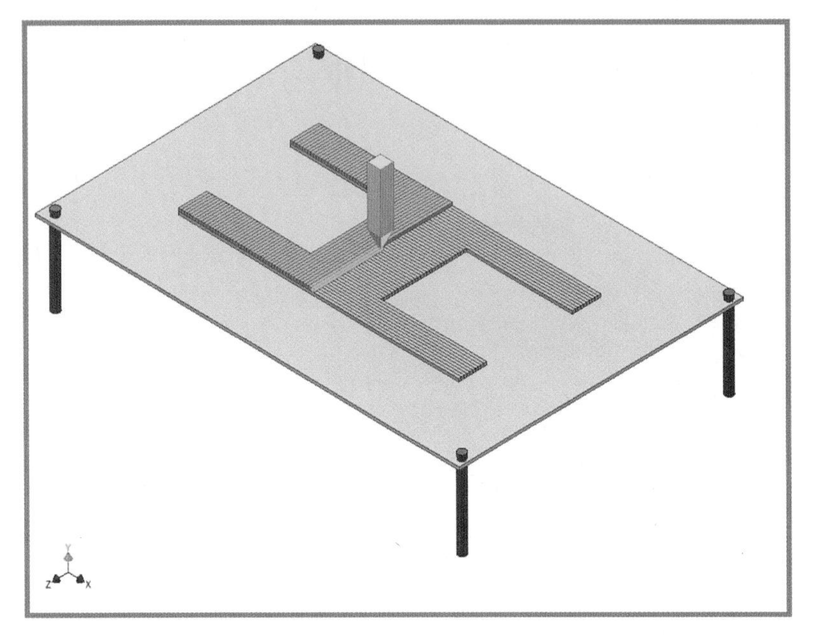

FIGURE 2.21 Layers are usually printed with alternating alignments to prevent possible "grain" effects.

A More Realistic 3D Printing Example

The previous 3D printing example was pretty trivial. So, let's look at Figure 2.22 that uses the mug shown in Figure 2.1 to illustrate a more realistic example.

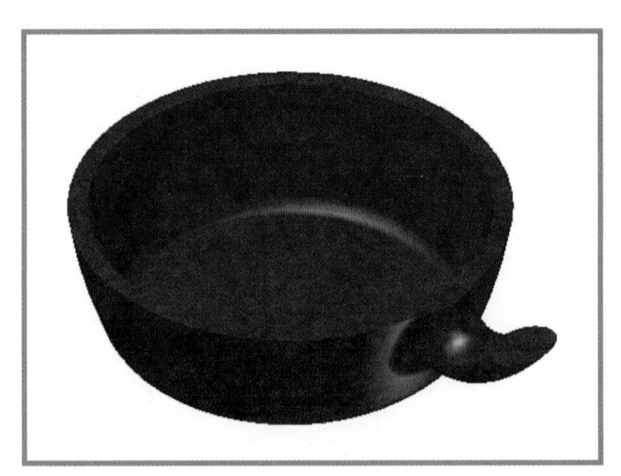

FIGURE 2.22 We're partly finished printing a more realistic example.

Figure 2.23 shows the contents of the next layer in isolation.

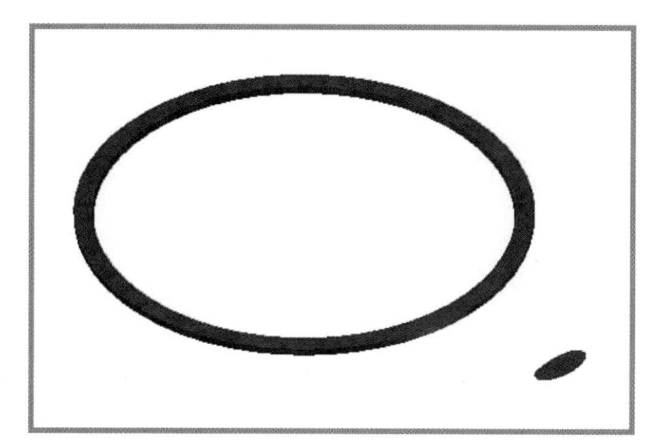

FIGURE 2.23 The next printed layer is shown in isolation.

In the real world, Figure 2.23 would look almost exactly like Figure 2.22 except that the part would be thicker by the thickness of the layer shown in Figure 2.22. I figure it's about time for a coffee break, and so Figure 2.24 shows the finished product.

FIGURE 2.24 Coffee, anyone?

As you have seen, 3D printing is effectively the opposite of sliced bread, but that doesn't make it the worst invention of all time. Far from it.

THROUGH THICK AND THIN

Earlier, I made reference to "hair's width" and "paper-thin." So what are the real numbers? A human hair is about 0.002" thick (commonly pronounced as *two thou*) or 0.05 millimetres, while bond writing paper is about 0.005"/0.13mm thick.

To get a rough idea of what this means, look at the edge of a stack of freshly opened printer paper to see 0.005"/0.13mm resolution.

By comparison, 3D printers typically have similar resolutions—in the range of 0.010"/0.25mm down to 0.001"/0.02mm. The finer the resolution, the more expensive the machine will be and the longer it will take to produce models, but the better the surface finish and the smaller the detail that can be produced.

So, how can you tell the true color of a person's hair? Measure it. Black hairs are about 0.0022", brown/brunettes are about 0.0020", and blondes are about 0.0018".

Two Basic Types of Printers

There are two kinds of 3D printer in the world. No, this isn't the lead-in to a bad joke. There really are two basic types of 3D printer: deposition and fusion.

Deposition—Deposit This...

In of the ***deposition*** printer, material is deposited a stripe at a time, layer upon layer, as implied by the previous sections. The material is usually one of the many types of thermoplastic (that is, "meltable" plastic) and is supplied in the form of a spool of plastic filament, much like weed trimmer string. It is fed through a heated print head where it melts and is extruded onto the part much like glue coming from a hot-melt gun. The popular MakerBot machine is an example of this type.

NOTE

The extreme example of deposition printing is a truck-mounted unit that can 3D print a full-size concrete house.

Fusion—Take This Material And Stick It...

The other type of printer uses fusion. These printers start with a thin layer of material, either as a liquid or fine powder, that covers the entire plate. The print head traverses over the layer fusing the material into solid as appropriate. The layer is lowered, fresh material applied over the entire printable area, and the process repeats.

How do you carve a statue of an elephant? You get a big block of marble and cut off anything that doesn't look like an elephant. How do you 3D fusion print a statue of an elephant? You get a big pile of loose raw material and then stick together everything that *does* look like an elephant.

When you've finished printing, the part is fished out of the unused material; the unused material can be reused for the next part.

Fusion printers come in two basic variants. One variant uses a liquid plastic that solidifies where the ultraviolet laser in the print head shines on it. The first commercially successful 3D printers were the StereoLithography machines from 3D Systems, which use the fusion system.

The other variant of fusion machines uses powdered materials. This can range from corn starch, which is glued together using sugar water from the print head, up to steel powder that is fused together using a high-powered laser. A number of brands and models are available.

Stacking Up: The Third Type of 3D Printing Process

Okay, I lied. There is a third type of 3D printing process, but it's quite rare. It cuts the individual layers out of a sheet material; then the layers are stacked and glued together into the final part. The good news is that you don't even need a special 3D printing machine. You can simply print the individual layers one at a time onto ordinary paper or thin cardboard and cut them out with scissors. The bad news is that it takes a lot of time and patience to stack and glue together the individual sheets for even a small, simple model.

Pros and Cons of 3D Printing Processes

In this section, you learn about some of the advantages and disadvantages of each process type.

Pros and Cons of Deposition Printers

First, the good news:

- Relatively inexpensive home-hobbyist machines such as the MakerBot are readily available.
- They can use a wide range of thermoplastic (meltable plastic) materials within the same machine.
- Parts usually come out ready to use with little or no cleanup required.
- Hollow, sealed cavities can be made, such as a basketball.

Now the bad news:

- Parts might require temporary support structures that need to be removed later.

Consider the coffee mug shown earlier. What if we want the handle to be open at the bottom, as shown in Figure 2.25, instead of curving back into and joining the body of the mug?

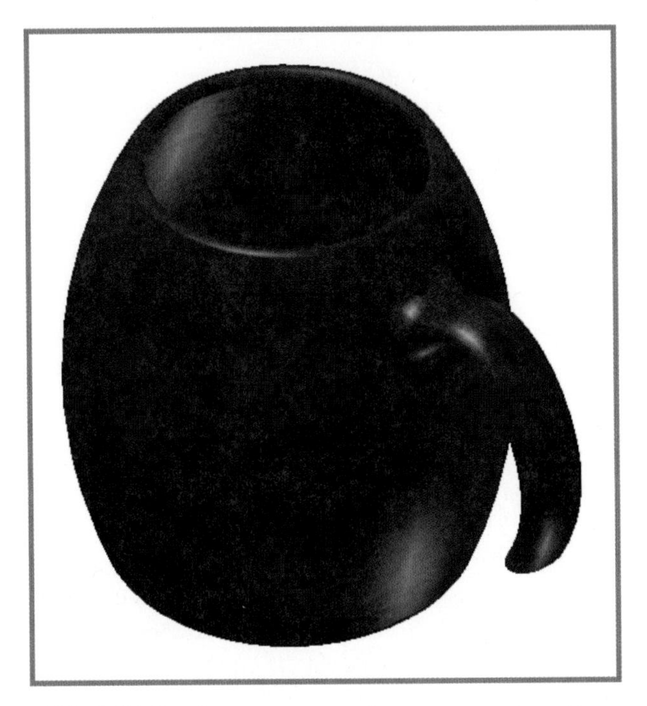

FIGURE 2.25 We want to print an open-handled mug.

Figure 2.26 shows everything starting out much as we expect it would.

FIGURE 2.26 Our printing process starts off quite normally.

Figure 2.27 shows everything looking good so far...until we reach the bottom of the handle.

FIGURE 2.27 MakerBot, we have a problem!

As shown in Figure 2.28, there's nothing to support the first bit of the end of the handle. Figure 2.28 shows how the first layer of the lower end of the handle would simply fall down because we are trying to print on thin air.

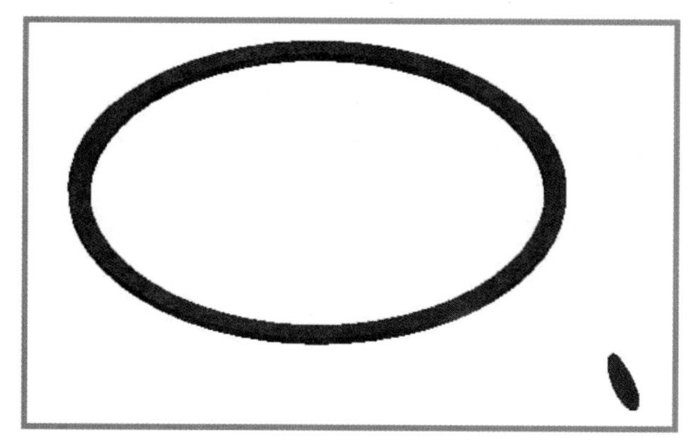

FIGURE 2.28 The tip of the handle has nothing to support it.

This portion of all subsequent layers will do likewise until we reach the point where the upper end of the handle curves back and joins the main body.

Looking back at Figure 2.23, it seems like it would have the same problem. However, there is a difference. Figure 2.23 shows that the next slice being printed is supported by the lower portion of the handle. Now look at Figure 2.27, which doesn't have the lower handle portion.

There are three possible solutions:

- Redesign the part to eliminate the overhangs.
- Add temporary support webs that are removed when the part is finished.

TIP

The good news is that the support software for most 3D printers is able to automatically add the kind of temporary supports needed in cases like Figure 2.27. The supports then must be removed manually when the 3D print is complete.

- Reorient the part for printing. In the case of the coffee mug, you could simply rotate the computer model 90° on its Y axis so it's lying on its side or 180° on its Z axis so it's upside-down.

Pros and Cons of Fusion Printers

First the good news:

- You can usually ignore the overhang problem because the unfused material will support things until the print operation is finished. This makes it easier to design and print complex shapes.
- A wide range of materials can be used from liquid plastic to corn starch to powdered metal.

Now the bad news:

- You can't build empty sealed cavities such as a basketball because they will be full of unfused material. If you need such an item, then you'll have to leave a drain hole that is plugged later.
- Any given machine can use only a specific material or narrow range of materials. You can't run powdered aluminum through a starch machine, for example.
- Parts need to be cleaned after printing. Machines that use a fine powder—starch in particular—tend to produce a lot of dust.
- Fusion printers tend to be more expensive than deposition ones.

Later chapters in this book provide more specific information on the pros and cons of the various 3D printers, but this gets you off to a good start.

Summary

As indicated in the opening paragraphs, 3D printers work from 3D computer models. How do you get the model? You can download existing models from websites, or you can create your own. This chapter described the basic operating principle used by all 3D printers and then went on to cover the two main types of 3D printers.

The following chapters explain how to create 3D models with the iPad, a Mac, or a PC.

123D Creature for iPad

A great starting point on the path to 3D modeling is the free iPad application called 123D Creature. It allows anyone with an iPad to use their fingers to create creatures of all shapes and sizes on the tablet's touchscreen.

→ *A free copy of 123D Creature for iPad can be downloaded from http://www.123dapp.com/creature, or it can be found on the Apple App Store on your iPad.*

This is a fun and easy way to create a 3D model of the monster out of your wildest imagination. When you've finished your masterpiece with Creature, you can export it from the iPad to print on your own 3D printer or send it to a service that will 3D print it for you.

Don't worry if you don't think you have any artistic ability because a large catalog of creatures is available online that you can use as a starting point and make them your own.

Let's begin by launching the application on the iPad (see Figure 3.1).

FIGURE 3.1 123D Creature (iPad).

Creating a Creature

When you first launch the Creature application, you are presented with three boxes: Create New, My Projects, and Community:

- **Create New**—This lets you start from a basic skeleton model that you will embellish with your own ideas using a number of steps along the way.
- **My Projects**—Shown in Figure 3.2, this is where you can view previously created works. You can store your creatures on the iPad or on the Autodesk servers, represented by the cloud icon on the right.

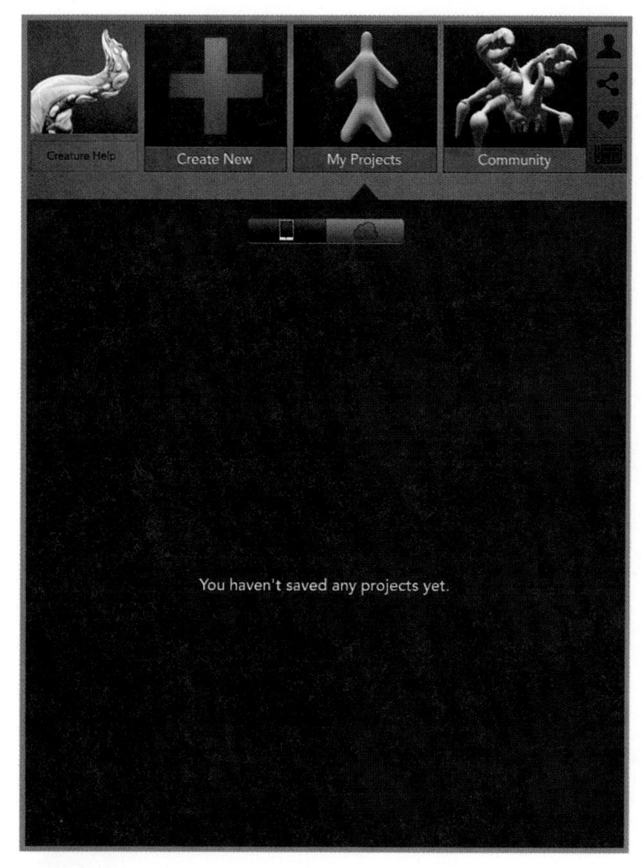

FIGURE 3.2 My Projects.

TIP

If you save your projects to the cloud, you can then access them from other Autodesk applications and further manipulate them.

- **Community**—Shown in Figure 3.3, this is a collection of all the publically shared Creature models that others have designed on their iPads and made freely available. Search has a number of options including popularity. Entering a keyword will filter the Community creatures by their tags.

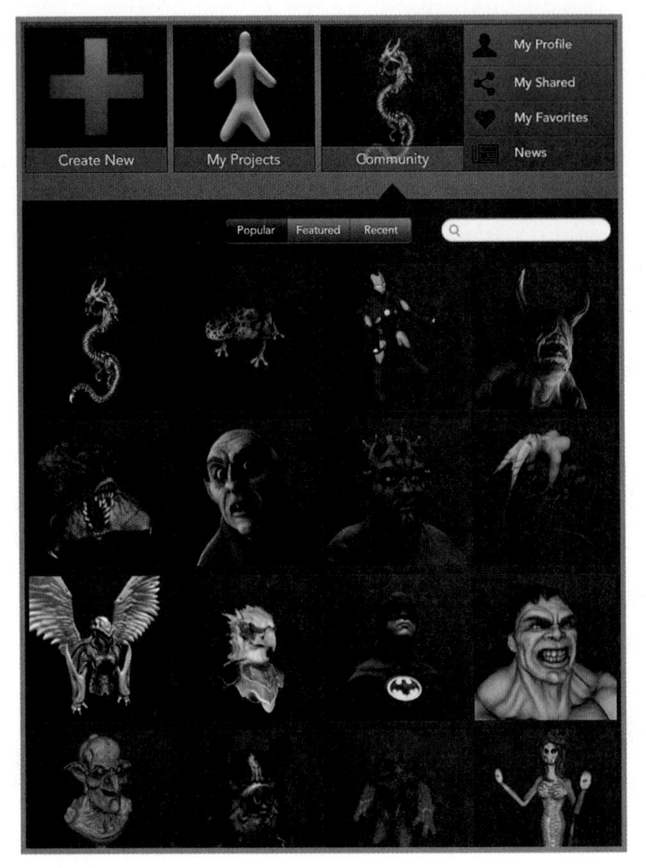

FIGURE 3.3 Community Creatures.

NOTE

While browsing the models, you can flag your favorites and copy them to your My Projects section to work on yourself or share them with friends on various social networking sites. A My Profile section shows your Autodesk profile and any shared Creatures you've made and shared.

- The last option in the Community section is for news about the application and featured designs from Autodesk (see Figure 3.4).

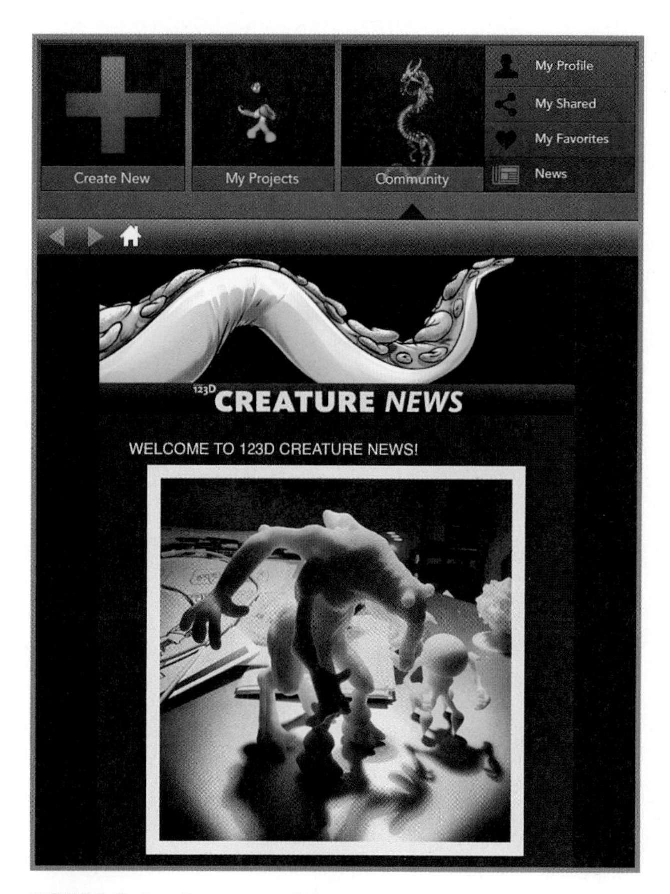

FIGURE 3.4 Creature News.

When you click the Create New box for the first time, you might be presented with a set of introduction screens (see Figure 3.5). These screens give you a basic overview for using the software and the steps your creature design will go through.

FIGURE 3.5 123D Creature welcome message.

Adding Bones and Joints

All creatures created with this program start with a basic skeleton. By adding joints, bones, and limbs (see Figure 3.6), you build up your creature's internal structure and general body shape. This structure allows you to create a creature and then pose it in various positions, just like an action figure, without having to redraw it each time.

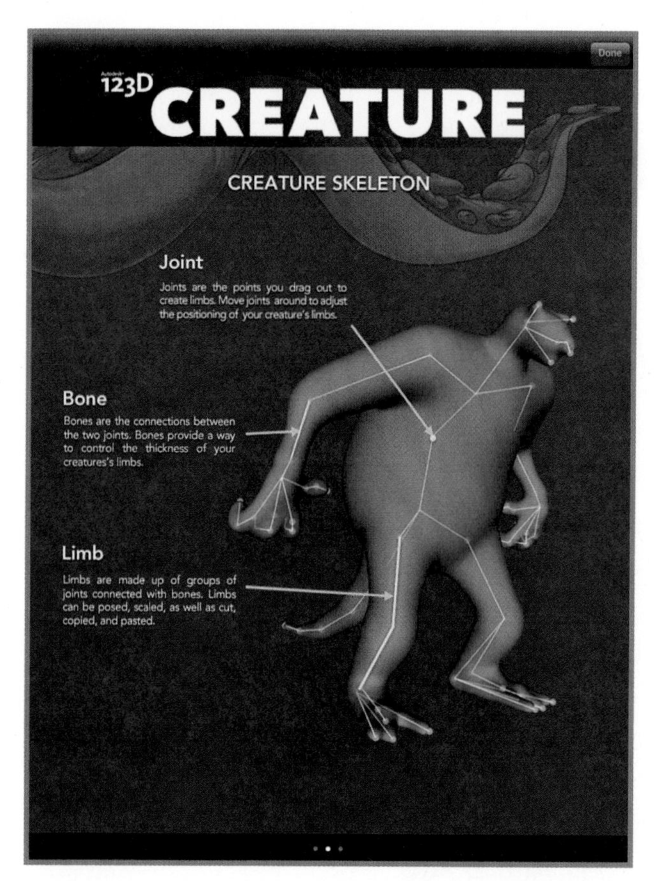

FIGURE 3.6 Creature skeleton composition.

The 123D Creature understands the relationship between bones and joints and will allow you to bend a leg. However, it will only bend as far as the joints and bones will allow, just like a person. *Limbs* are a grouping of the bones and joints to make it easier to move a larger segment of your creature.

Moving the Model

During creation, you'll need to regularly move your model around. To do this, Creature enables you to move the model in a number of ways (see Figure 3.7):

- **Tumble**—The way to rotate the model in any direction by using one finger that is not touching the model anywhere. Be sure you press down on to any area in the background, as long as it's not your model. Doing this rotates the model in the direction you drag your finger along the touch screen.
- **Zoom**—Allows you to zoom in or out on your model using two fingers in an expanding or a contracting action.

- **Pan**—Requires two fingers and allows you to move the model around 3D space without rotating it.

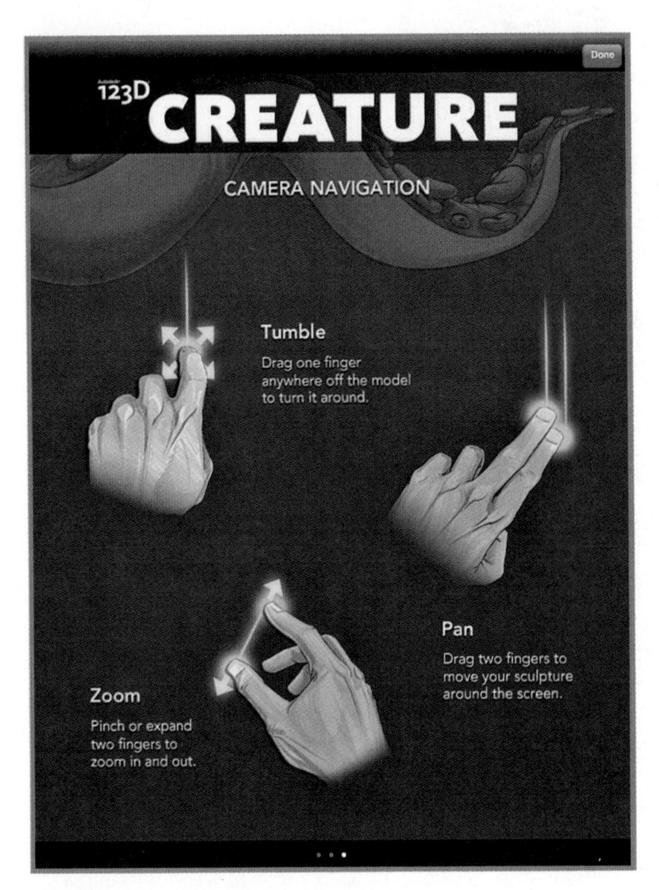

FIGURE 3.7 Camera navigation.

Sculpting Your Creature

After you've finished designing the skeleton, you'll then go to the sculpting area to further enhance your model with various tools (see Figure 3.8). Think of this part of the process as if you're working with a clay model, allowing you to push and pull various parts of the virtual clay to shape your creature.

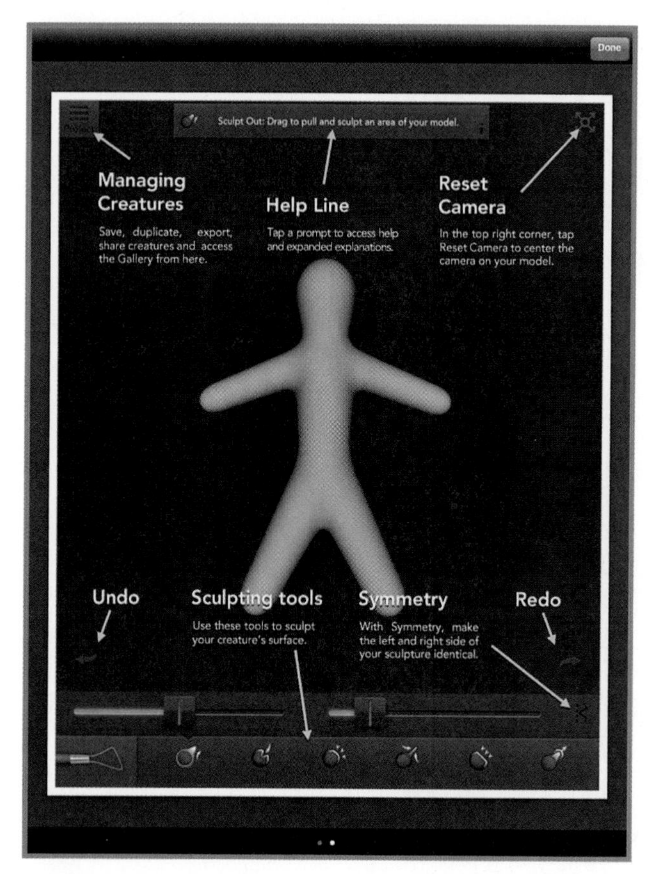

FIGURE 3.8 Getting around the main sculpting window.

To simplify the design process, each creature is essentially made as a mirrored model. This means you can modify the left side and the right side will do the same thing. You can turn off this feature called Symmetry in the sculpting section on the lower-right corner (see Figure 3.9).

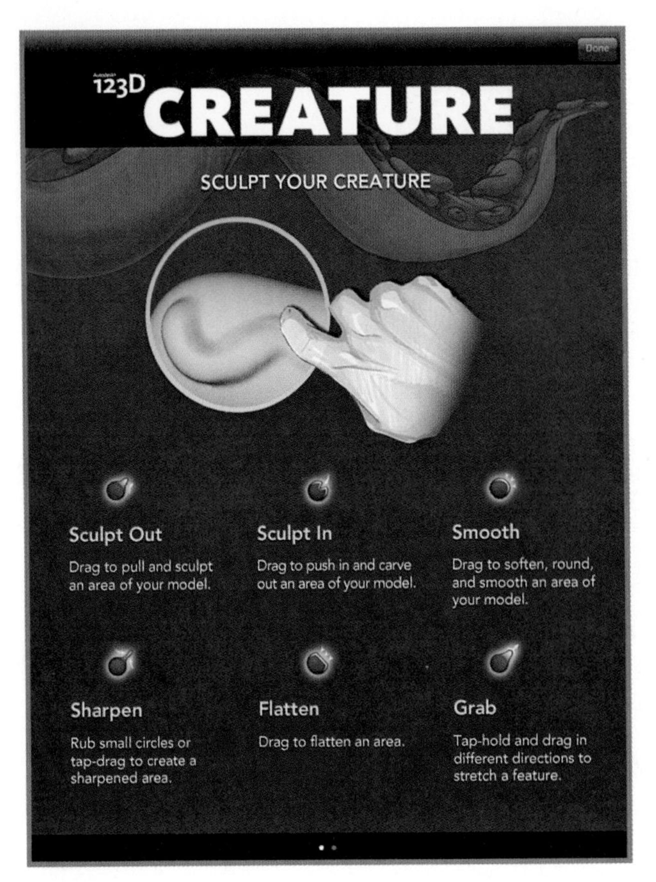

FIGURE 3.9 Sculpting controls.

Let's start making something scary...or possibly just weird and creepy.

Working with Control Options

After creating a new skeleton, you are presented with a basic skeleton shape that kind of looks like a simple gingerbread man (see Figure 3.10). Try some of the previously mentioned camera controls to get a feel of how the interface works before you start creating your creature. Fortunately, there is an Undo button on the lower left of the screen in case you make a mistake while learning to navigate.

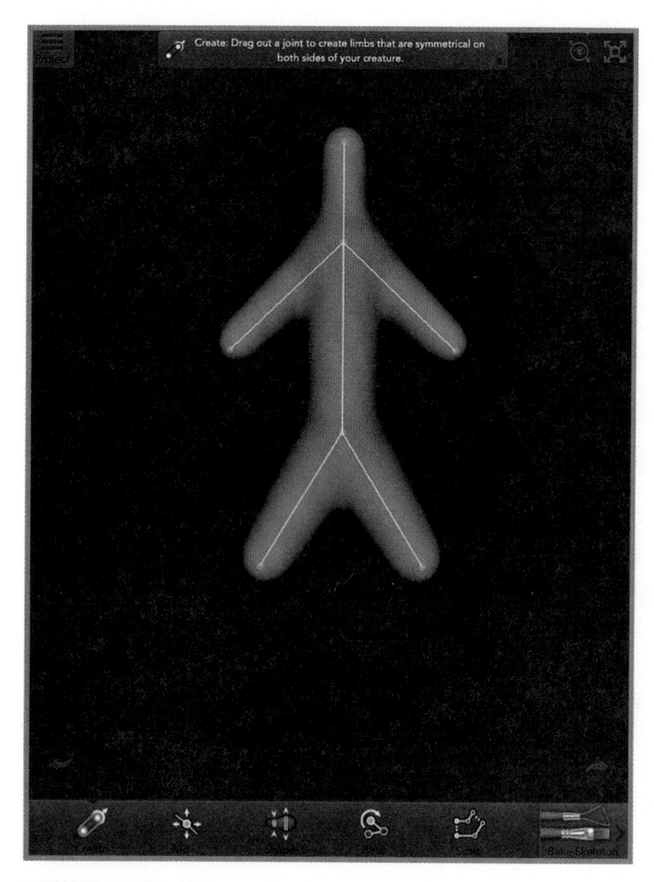

FIGURE 3.10 Sculpting controls.

If you wind up with a view that doesn't work for you, you can press the Reset Camera button in the upper-right corner of the screen (see Figure 3.11).

FIGURE 3.11 Resetting the camera.

A number of controls listed in Table 3.1 can be used while manipulating your skeleton (see Figure 3.12).

TABLE 3.1 Skeleton Controls

Create	Drag out a joint (the blue dot) to create new limbs that are symmetrical on both sides of your creature.	
Move	Drag a joint (blue dot) around to reposition it. If you need to access the back of the model, tumble the camera by pressing on any part of the screen that isn't the model to rotate your view.	
Shape	Drag up or down on a blue bar to thicken or thin a bone.	

Pose	Press and drag on a joint (blue dot) to pose the model in a different direction. Tumble the camera to rotate in a different direction.	
Scale	Press and drag on a joint (blue dot) to enlarge or shrink a limb.	

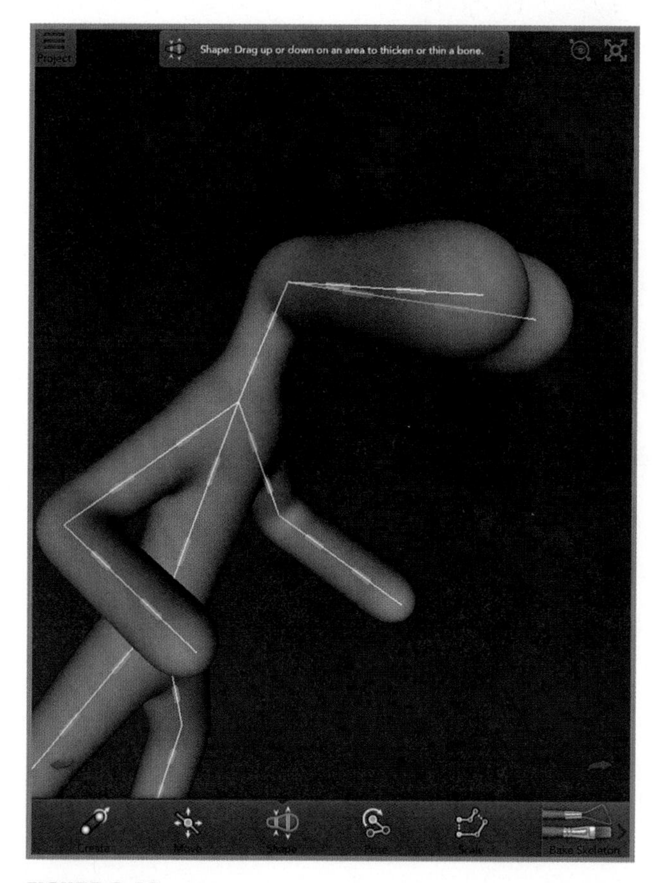

FIGURE 3.12 The creature is starting to take shape.

The last icon, Bake Skeleton, will take you to the next step in the process. For now, however, use the options in Table 3.1 and let your imagination run wild (see Figure 3.13).

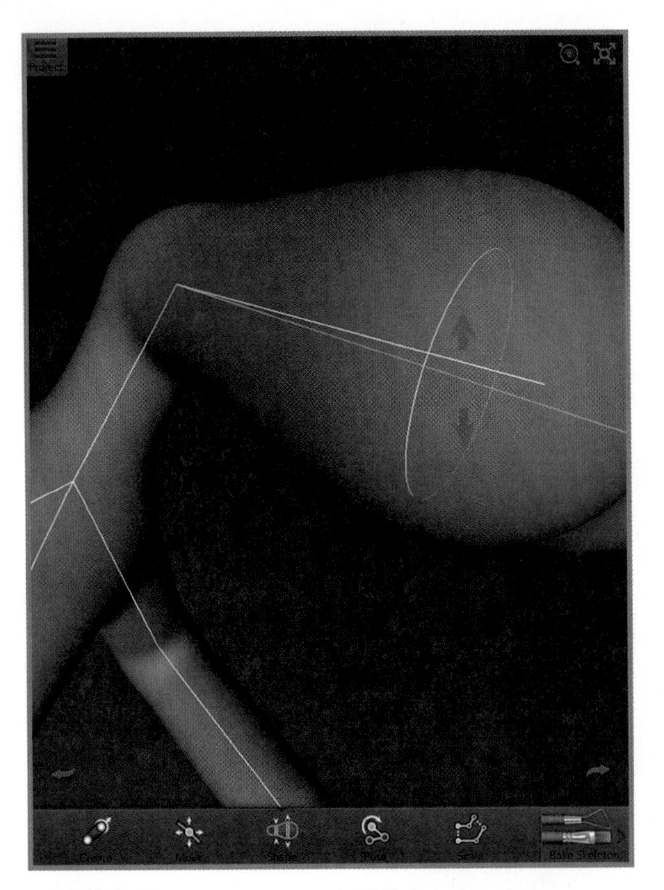

FIGURE 3.13 Actions are highlighted in blue.

As you activate various control options, the parts you can manipulate are highlighted in blue within your model. If that action has multiple axes that can be manipulated, you'll see colored circles with directional arrows, as in Figure 3.14. These indicate which axis (X, Y, or Z) you are manipulating. You might want to tumble the camera to better see the movement angles.

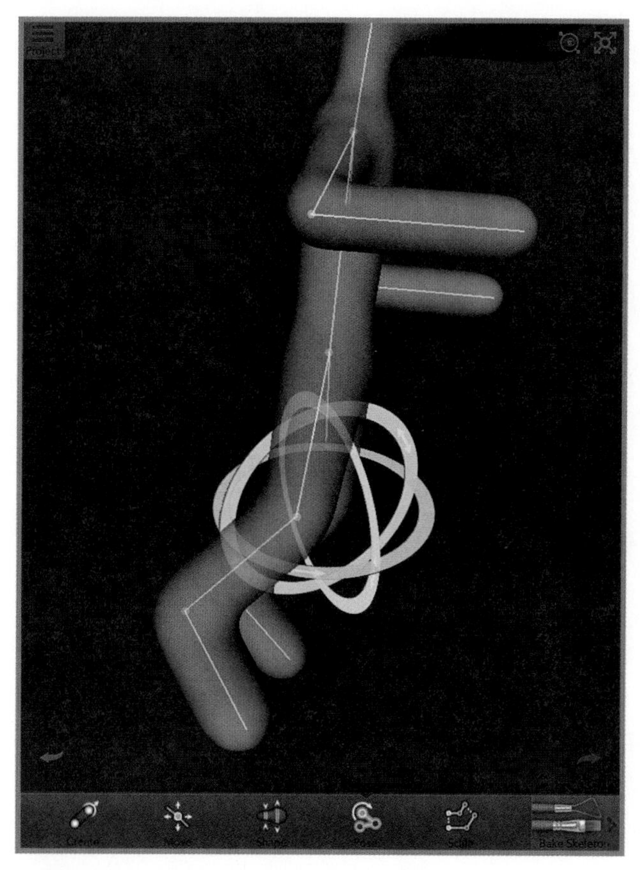

FIGURE 3.14 Multiple axis controls.

When you're happy with the skeleton you've created, press the Bake Skeleton button on the bottom right to go to the Sculpt section after the baking process (see Figure 3.15).

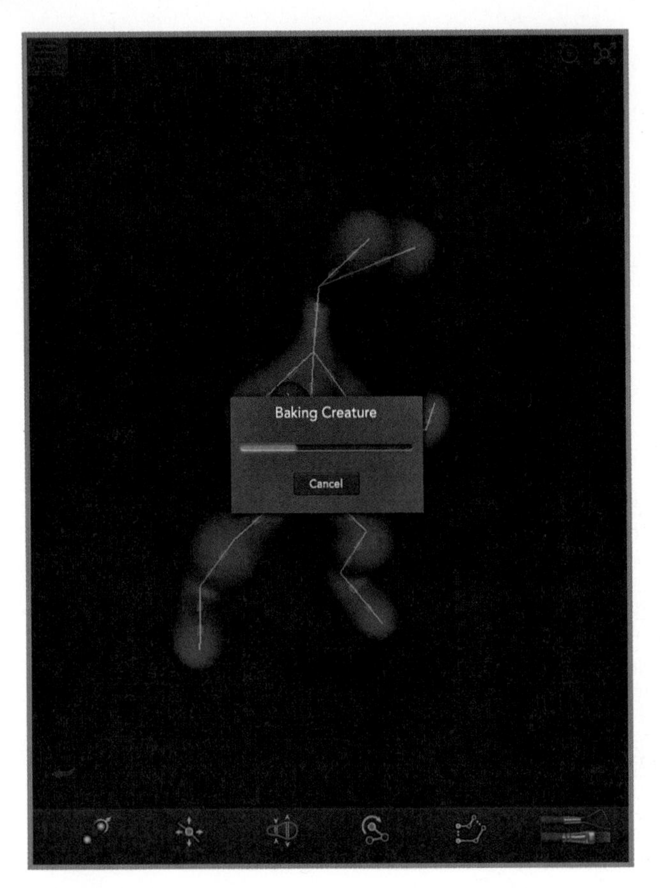

FIGURE 3.15 Baking your skeleton.

As shown in Figure 3.16, you're ready to start detailing your creature.

FIGURE 3.16 Ready for detail work.

TIP

The Sculpt section contains a number of options that affect the physical appearance of your creature, while others are superficial image effects that just present the creature in a dramatic way.

NOTE

If you're planning on 3D printing your model, you might want to skip a few of the image effects options (like lighting and backdrops) because they may not be printed on in your final 3D model, which is a physical object. This can vary depending on the type of 3D printer you have access to or if you're sending your creature to a third-party printing service that can 3D print in full color.

While sculpting, you can vary the size and strength of the various tools' brushes using the blue sliders on the bottom of the screen.

Note the blue line in the middle of the creature in Figure 3.17, which is the dividing line when Symmetry is turned on. Anything done on one side of the model will be reflected on the other side as long as this is turned on.

FIGURE 3.17 Sculpting your creature.

Don't forget to look at the back of your model as well while you're sculpting (Figure 3.18)! This also applies to painting, which as mentioned earlier, can have little effect on your model if you're 3D printing it in most cases.

FIGURE 3.18 Get around to all sides of your creature while painting to ensure complete coverage.

Notice in Figure 3.19 that adding backgrounds and lighting to your model has little effect on the actual 3D model. However, it can be fun to play with these options.

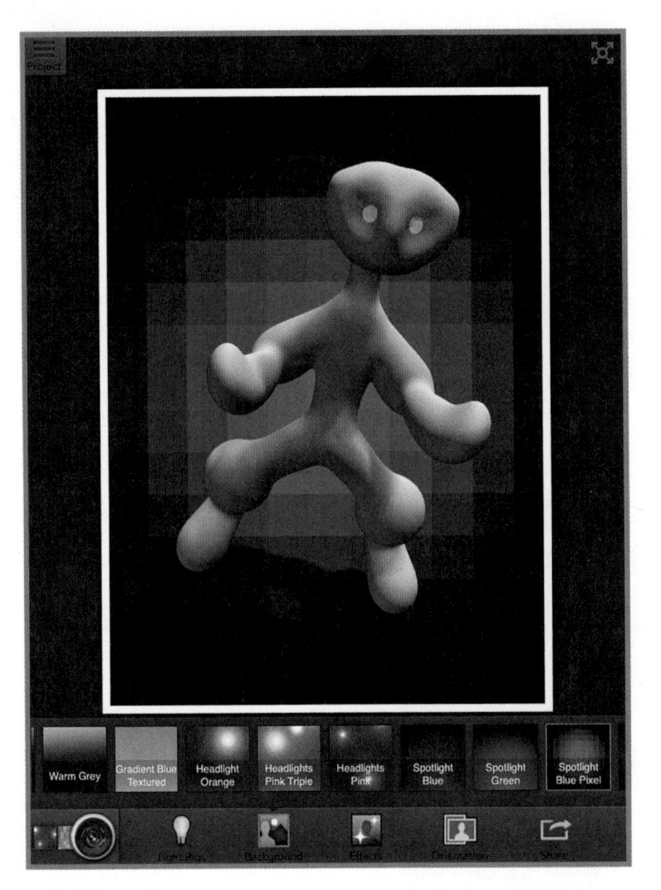

FIGURE 3.19 Effect of backgrounds and lighting.

There are a lot of image options to play with that can really add lots of great, if not creepy, detail to your model (see Figure 3.20). It's easy to spend a lot of time trying the various options.

FIGURE 3.20 OMG! What have I done?

Sharing Your Creature

When you're happy with your creature, press the share button on the bottom right (see Figure 3.21). From here, you can send a photo to various social networks or save it on your device.

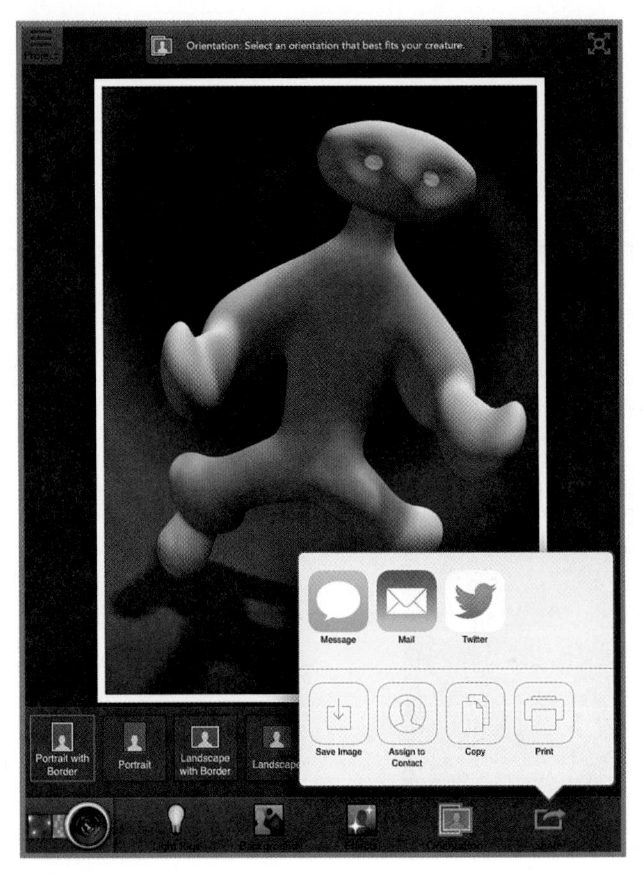

FIGURE 3.21 Sharing your creature.

You can save your creature by selecting the drop-down on the upper left of the screen (see Figure 3.22).

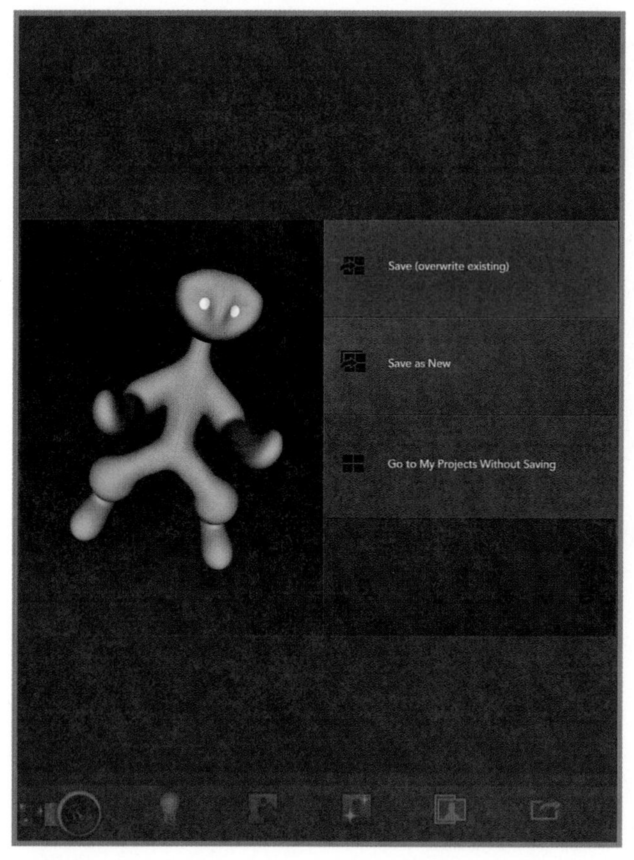

FIGURE 3.22 Don't forget to save your creature.

After you save, you'll get another menu of options to make a duplicate, export your creature as a 3D mesh, order a 3D print, or share your creature with the community (see Figure 3.23).

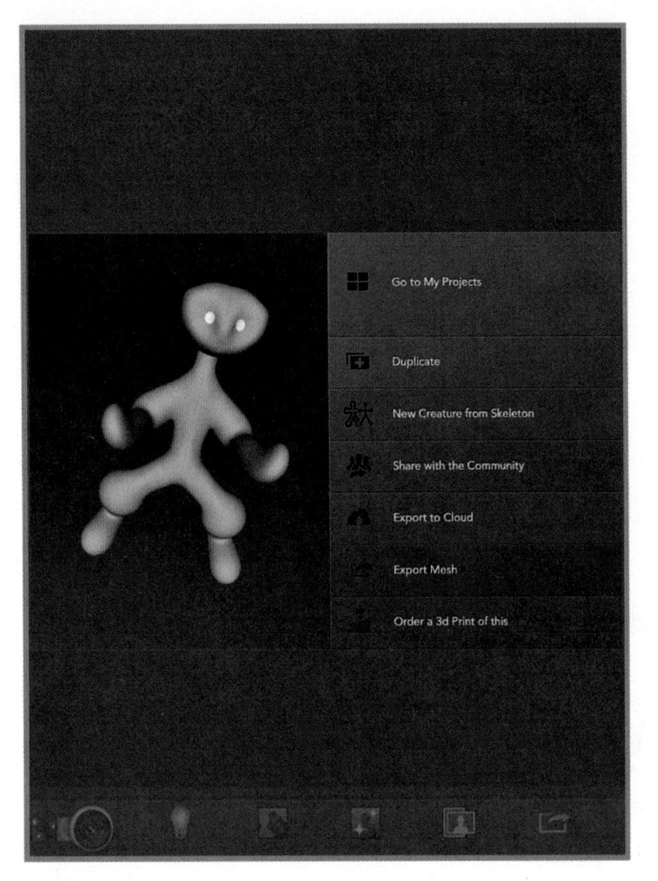

FIGURE 3.23 Options after saving.

You can either send your exported mesh file via email or send it to iTunes so you can retrieve the file from within iTunes (see Figure 3.24). If you choose email, it will create a compressed Zip file and launch your email program with the file as an attachment.

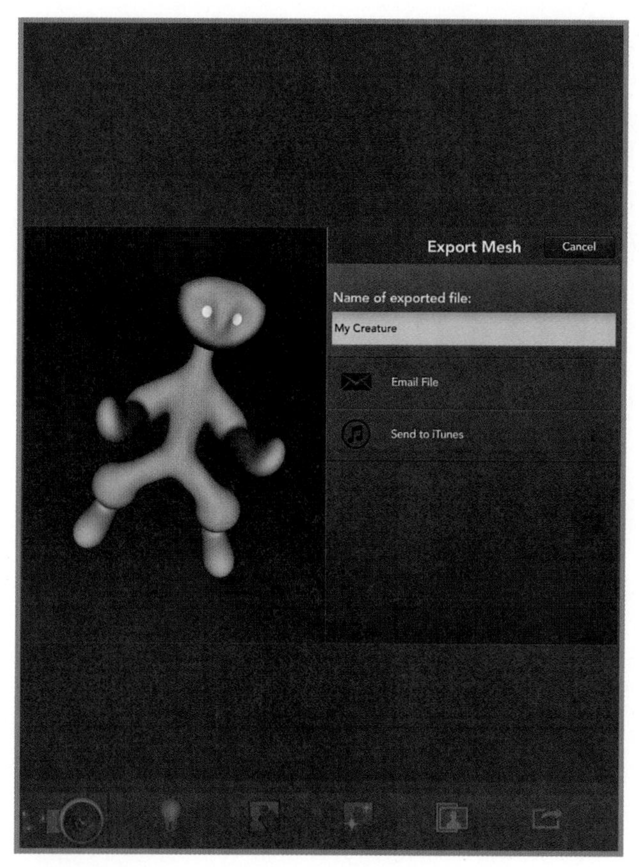

FIGURE 3.24 Exporting your mesh.

If you choose to order a 3D print of your creature, you'll be taken through the steps to order from Sculpteo right within the Creatures app (see Figure 3.25).

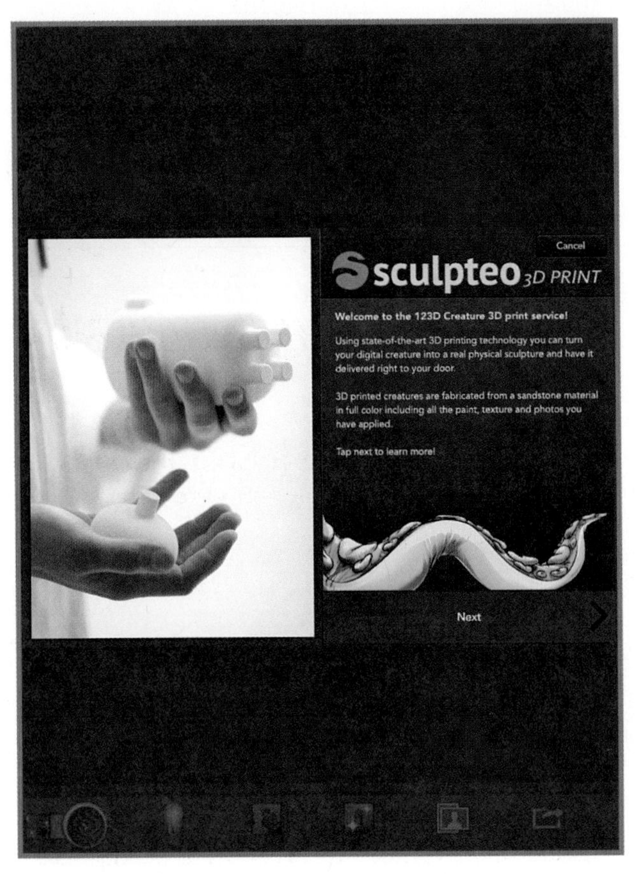

FIGURE 3.25 Ordering a 3D print of your creature from Sculpteo.

→ *To learn more about working with a 3D printing service like Sculpteo, go to Chapter 16, "Using a Third-Party 3D Printing Service Bureau."*

Summary

In this chapter we covered using 123D Creature for Apple devices to create monsters that can be as wild as you can imagine. Some features of the application such as lighting and textures can also be skipped if you're planning on only 3D printing your creature because they are purely for displaying your creation on your device.

Happy creature making!

Creating 3D Objects with Cameras and 123D Catch

Autodesk has a free service called 123D Catch that allows you to scan objects in 3D of just about anything, including people or buildings, using a regular digital camera—even with the one in your smartphone or tablet.

NOTE

You will need to set up an Autodesk account (it's free) before you can start. The process is quick and only requires a valid email address. It can be done via the application directly or via the website at http://123dapp.com.

There are a couple of ways to use 123D Catch. In this chapter, we cover the use of the iOS and web versions of the software. Both require that you take multiple photos of your object, from every angle possible. Then you submit those photos to the 123D servers either from within the Apple iOS app or via your browser.

NOTE

123D Catch is not currently available for Android devices but works with any mobile iOS device from Apple that has a camera such as an iPhone, iPad (see Figure 4.1), or iPod Touch.

The 123D website then processes those images and presents you a 3D model of the object. We go over some tips to ensure the best possible success in capturing your model.

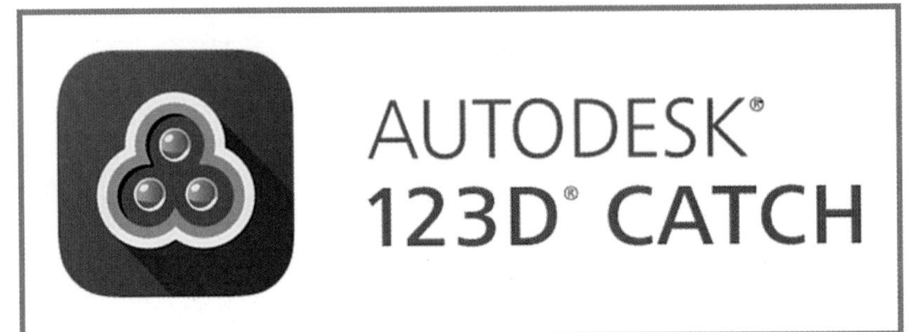

FIGURE 4.1 Launching 123D Catch application on iPad.

NOTE

123D Catch doesn't require the use of an Apple device. You can use any camera and upload the photos using the website version of the application. The same tips apply when using your own camera. The iOS app just saves you the step of having to manually upload the photos.

Photographing Objects

Let's start with a handmade ceramic cactus that was purchased at a street market in Mexico about 20 years ago. The scanning process works best with objects that aren't shiny because reflections can cause problems while processing the model.

TIP

If you have an object with a shiny or reflective surface, you can apply something such as cornstarch, flour, or a similar powder to make that surface less reflective. The same applies if you are scanning people. Dark hair doesn't register as well as lighter hair. Applying a white powder (such as cornstarch) to dark hair will assist in bringing out the details in the model when scanning.

Using an iPad and the 123D Catch app, approximately 30 photos of the cactus were taken from every angle of it (see Figure 4.2).

NOTE

The 123D Catch app is free from the Apple App Store. Visit http://123Dapp.com/ for more information.

FIGURE 4.2 Using an iPad to photograph the cactus.

Start at one side of the object, and work your way around it, keeping the object fully in the frame, approximately the same distance for each photo. Alternatively, you can set up your camera in a fixed position and rotate the object you're photographing. It's better to have some overlap of the object in each photo than not because the software will figure out where to stitch the images together better.

TIP

Ideally, the background in the photos should be high contrast or complex compared to your object. Scanning a solid red object on a red table for example will lead to a poor scan. Simply using a sheet of (black-and-white) newspaper underneath your object can greatly improve your scanning results. This contrast helps the software separate your object from the background in the photos.

You can review what you've captured by pressing the Review button, as shown in Figure 4.3. You can delete images or retake specific angles if they didn't turn out simply by tapping on the photo.

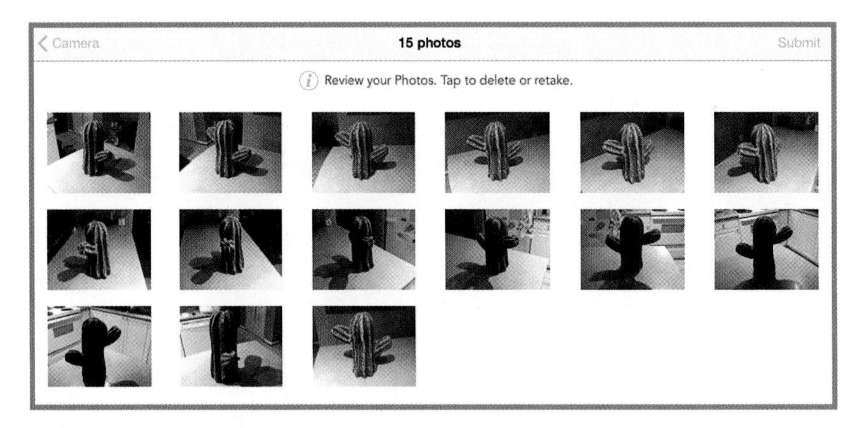

FIGURE 4.3 Gallery of captured images for review.

If you touch the screen on the iPad with one finger, you can rotate the model. Using two fingers at the same time, you can pinch and zoom to adjust your view of the model. The model also can appear upside-down depending on the orientation of the camera used to capture it.

When you're happy with the results, press the Finish Capture button, which takes you back to the Captures tab. Press the Tap to Process button to begin the process of uploading the photos to the Autodesk servers and begin converting the images into your 3D model.

NOTE

This process can vary in length depending on how busy the servers are and how complex your model is, but it usually takes only a few minutes.

When the software finished processing the model, it gave me the 3D model of the cactus (see Figure 4.4).

FIGURE 4.4 Reviewing the 3D model output from 123D Catch on the iPad.

FIGURE 4.5 Overhead view of 3D model.

Correcting the Model

There are a few issues with the model that need to be corrected; we walk through them here.

You'll notice that the top of the surface on which your object was sitting may also be included in the model. Shortly, we'll go through the steps to trim that off, leaving only the desired object behind as your 3D model.

After saving the model to your Autodesk account, you then can manipulate it further using the other Autodesk applications. We cover these in later chapters. Your model should appear in the Captures tab of 123D Catch on the iPad and in the My Projects section online after you log in to 123D Catch (see Figure 4.6).

FIGURE 4.6 Reviewing the finished output 3D model of the cactus in the online version of 123D Catch.

As mentioned, the bottom of the model contains the surface on which the object was sitting when you captured the images. We're going to use the Plane Cut tool to trim off the bottom of the model, leaving behind only the cactus object, as shown in Figure 4.7.

FIGURE 4.7 Plane Cut tool.

Select the Plane Cut tool from the toolbar. A set of controls should appear on your model with an arrow pointing along the Z axis. The bottom of this arrow represents the bottom of the model and direction the cut will occur. A circular control on both the X and Y axis lets you rotate the cutting surface and align it with the bottom of your model.

In Figure 4.8, the arrow is pointing down. It should be pointing up; otherwise the top of the cactus would get cut off. Click and drag your mouse on either the X or Y circle (below the purple line in Figure 4.8) to rotate the cut line. It should snap at the 45° and 90° positions to help with alignment.

FIGURE 4.8 Ensure the arrow is pointing in the correct direction of the desired cut.

When the arrow is orientated correctly, you can then click and drag it to move the cutting plane up and down your model (see Figure 4.9). In this example, you want to see only the purple line along the bottom of the cactus which means that the cut will discard the countertop surface.

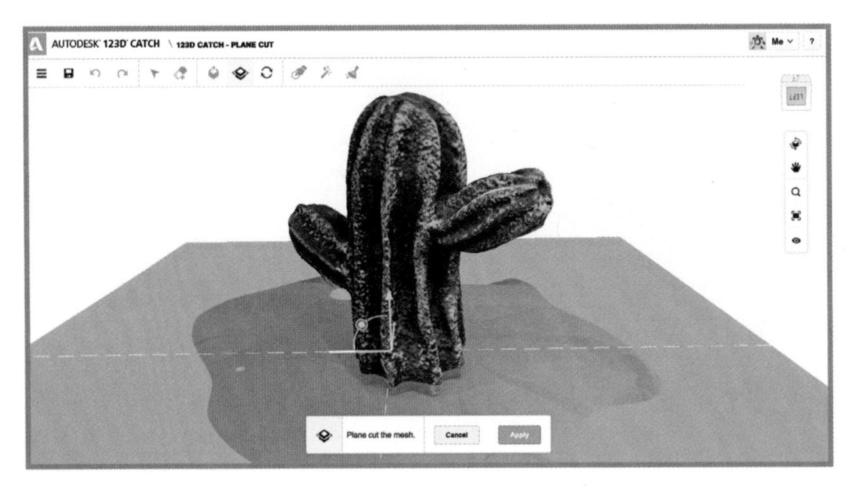

FIGURE 4.9 Moving the cutting plane.

Use the X- and Y-axis circles again to pivot the model along the plane until it's level. The purple line represents the new bottom of the model. Press the Apply button when you're happy with the position of the purple cutting line.

In Figure 4.10, you can see that the bottom of the model has been cut and the countertop surface has been removed from the model.

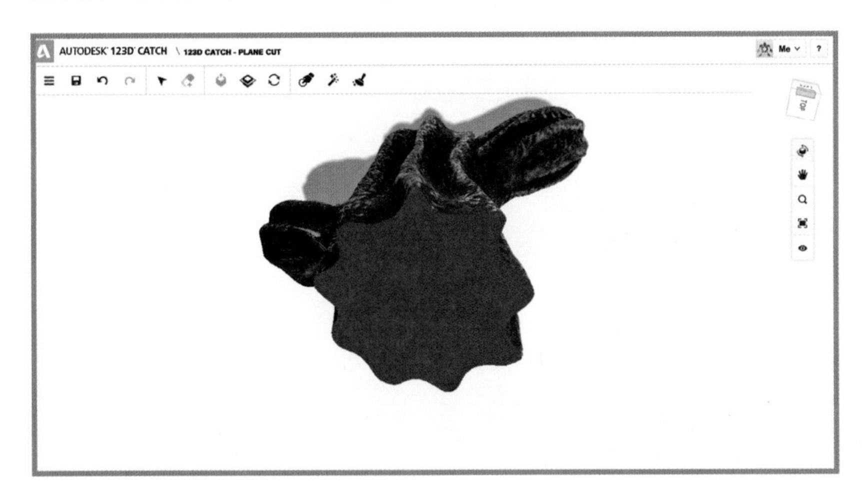

FIGURE 4.10 Cutting complete.

Orienting the Model

One last thing that you might need to do is orient the model so that it is upright for printing. You'll notice in Figure 4.11 that in the upper-right corner of the screen the orientation box is upside-down.

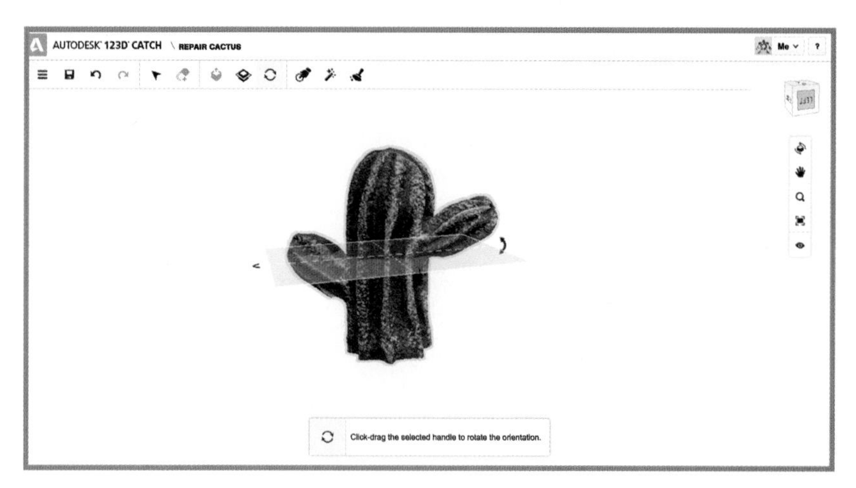

FIGURE 4.11 Orienting the model.

To correct this, press the Orientate button from the menu; a blue plane appears on the model. Clicking the arrow enables you to rotate the model and correct the orientation (see Figure 4.12). In this example you need to rotate the model 180°.

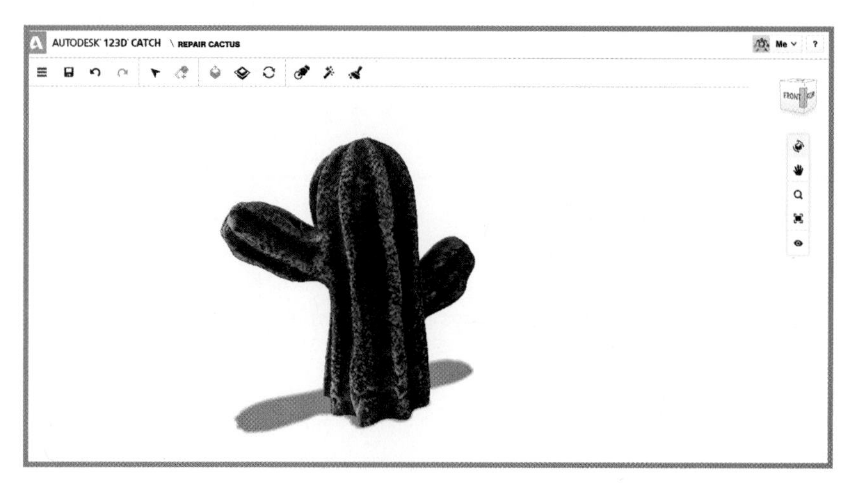

FIGURE 4.12 Model orientated correctly.

Note that the upper-right box is now upright. You'll need to save the model before you proceed to the next and final steps of repairing the model.

Repairing the Model

It's not uncommon for scans to result in some holes or rough spots due to the software misinterpreting your source photos. This can be caused by a number of reasons, but usually it's due to areas of the source object being in shadow or too similar to surrounding areas.

Fortunately, 123D Catch has some repair options included that will analyze the finished model. You can manually do some spot repairs to the model using the Smooth tool for rough areas and holes. Automatic repairs can be done using the Heal Mesh and Auto-Cleanup tools from the menu.

The Smooth tool is used to smooth out any rough points in the model that might not have been processed as accurately as desired. You can adjust the brush size using the slider on the bottom as well as the strength of the effect.

Start with small brush Size and low Strength; then work your way up (see Figure 4.13). You don't want to apply too strong of an effect because it will look obvious on the model. Also, be sure you rotate around the model to smooth all sides of the model.

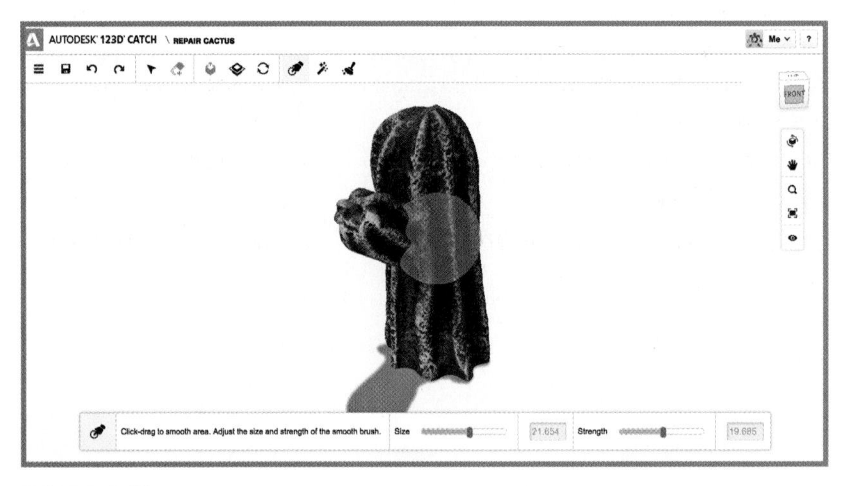

FIGURE 4.13 Smoothing the model.

The Heal Mesh tool detects any holes in the model and repairs them. Inspect the model after applying this to ensure it doesn't close openings that you intended to be open (see Figure 4.14).

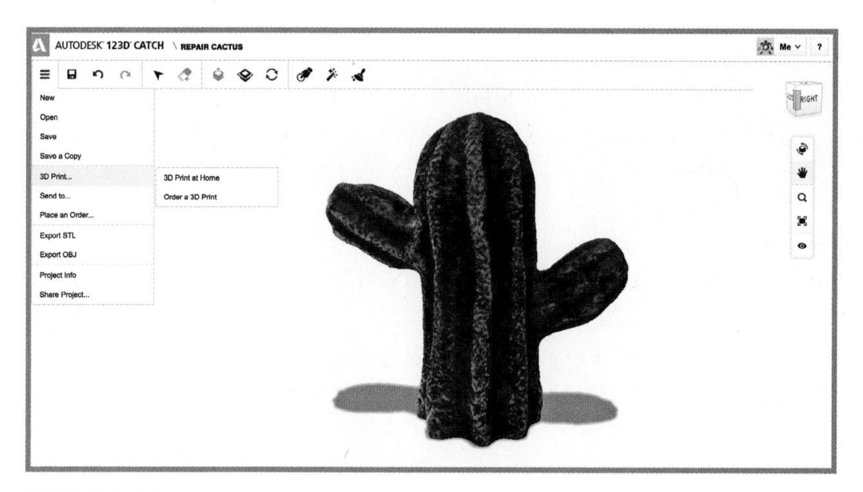

FIGURE 4.14 Getting ready to 3D print the model.

The Auto-Cleanup option magically fixes the model and removes any of the detached parts of the object. This is the last step you should apply. Again, inspect the model from all sides to ensure it didn't over-repair any parts of the model unnecessarily. Use the Undo option if you aren't happy with the results.

Finally, you can choose to preview the model for your 3D printer software or send it to a printing service (see Figure 4.15).

→ *For more information on how to send your model to a 3D printing service, go to Chapter 16, "Using a Third-Party 3D Printing Service Bureau."*

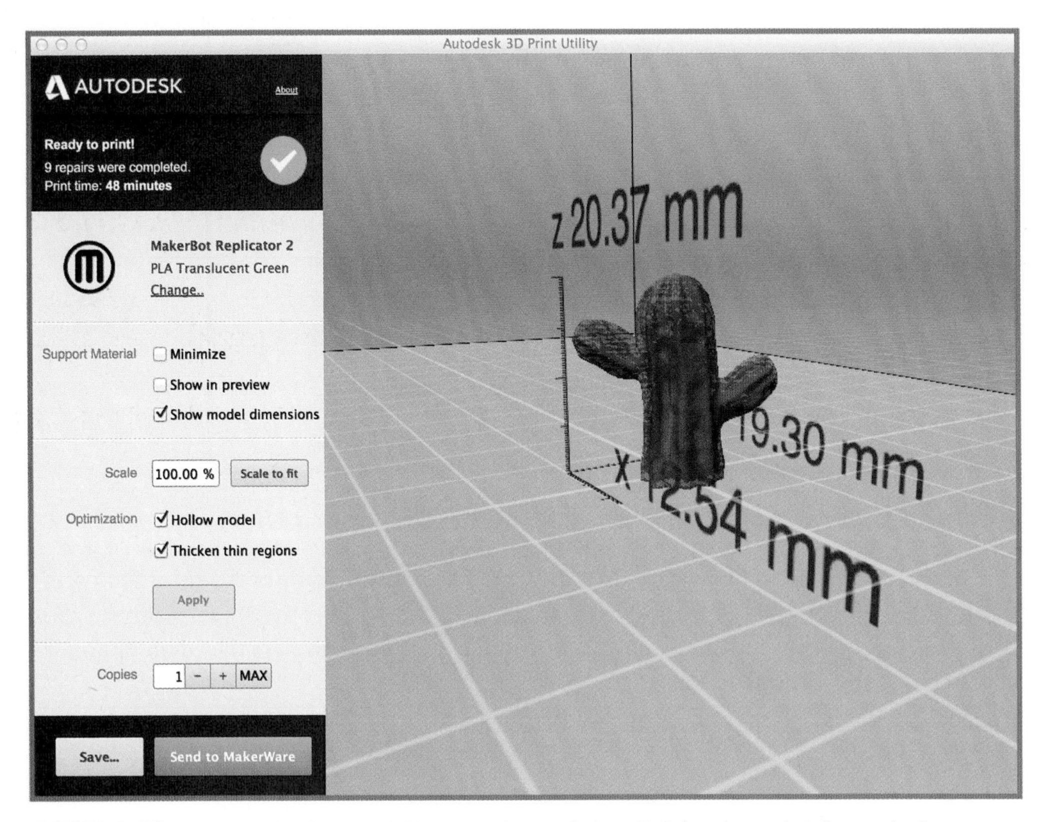

FIGURE 4.15 Autodesk Print Utility preview of the finished model for printing on a MakerBot.

Summary

In this chapter, we used an iPad to photograph a physical object and create a 3D model from those photographs. Then, using a number of built-in tools, we cleaned up and repaired the model, getting it ready to send to a 3D printer.

In the next chapter, we dive into creating models from scratch using your iPad and 123D Design.

5

Introducing 123D Design for iPad

123D Design for iPad is a free application available on the Apple App Store (see Figure 5.1). It's a great way to dive into design using the iPad's great touch interface.

FIGURE 5.1 123D Design for iPad.

When you first start 123D Design, you might be presented with a brief overview of the menus. If you swipe to the left through the overview, then you'll presented with a menu of choices along the left side, as shown in Figure 5.2.

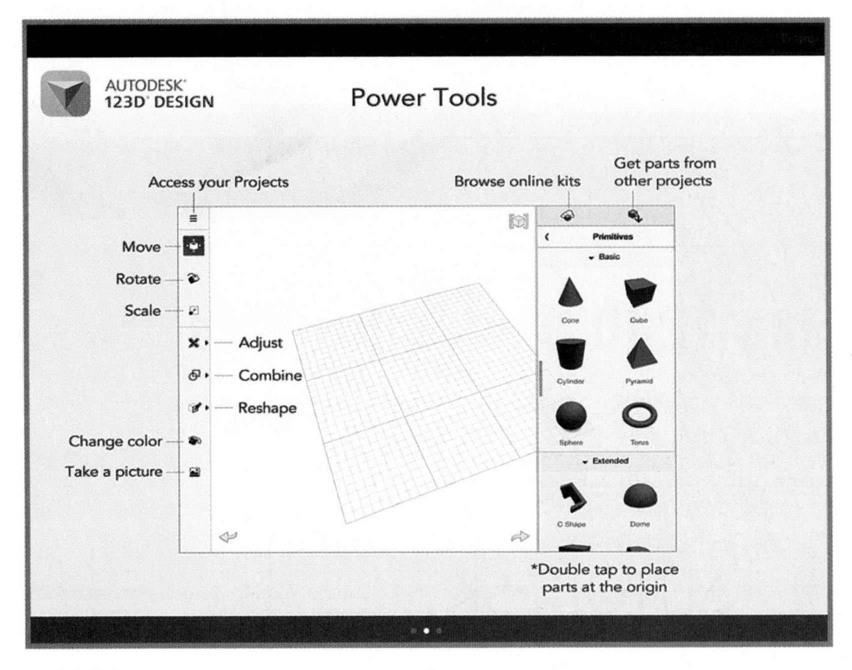

FIGURE 5.2 First launch of 123D Design.

The application comes bundled with a large array of sample designs and many parts you can incorporate into your own designs—everything from robots to battleships and parts of everything in between. You can find these designs in the Examples tab (see Figure 5.3) and various parts and objects in the Parts Library.

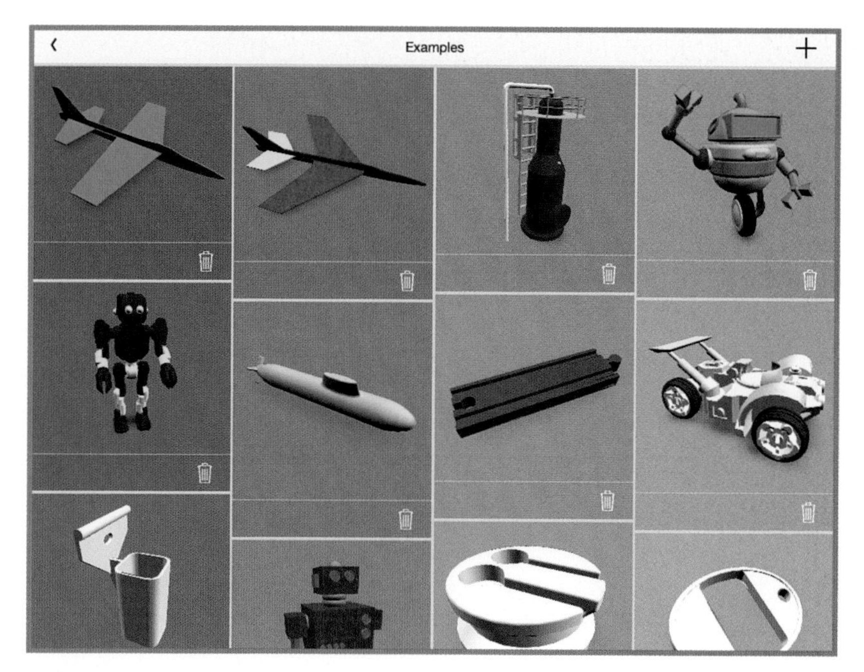

FIGURE 5.3 Examples of designs.

NOTE

Like other Autodesk applications, 123D Design for iPad works better if you have an Autodesk account (it's free to create one) so you can store and/or share your designs online. The account also makes it easier to move your designs from the iPad to the desktop application version of 123D Design because your designs will be available via the My Projects tab within both applications.

→ *A free online Autodesk account can be created at http://www.123dapp.com/. It takes only a minute, and you can sign in with any of the common social media logins or with just your email address.*

123D Design Interface

Let's take a tour of the interface to get familiar with it.

→ *In Chapter 6, "123D Design Exercises for iPad," we go through some design exercises.*

Selecting the New option from the upper-left menu takes you to the empty workspace, ready to begin your design work, as shown in Figure 5.4.

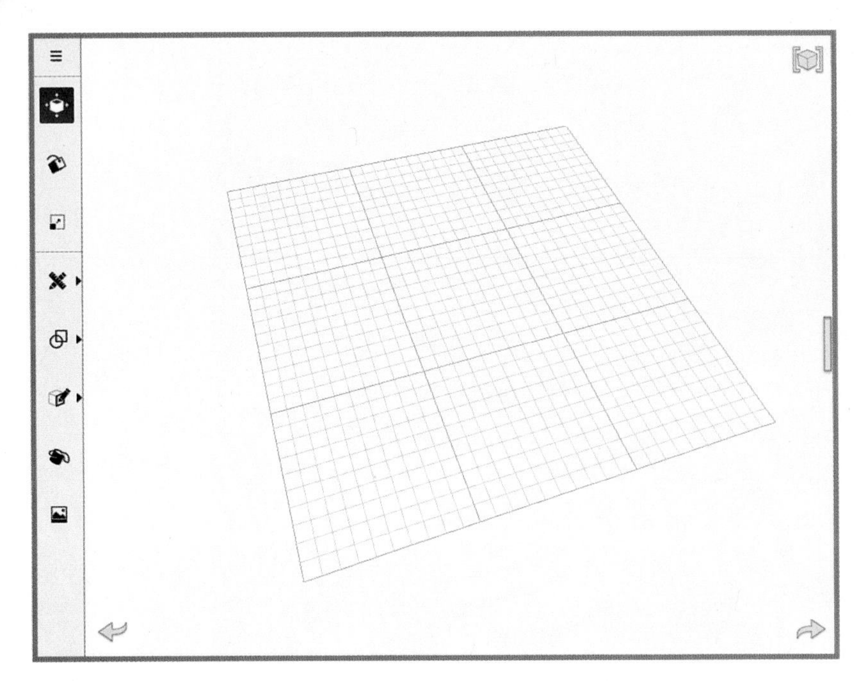

FIGURE 5.4 Blank canvas ready for designing.

On the right side is a menu of parts you can use in your design. These parts make starting with a basic shape or part and then resizing or reshaping it to your liking much easier. It also makes adding other parts to it easy.

Primitives Parts

The first category of parts is called Primitives (see Figure 5.5). It is organized into a number of subcategories. Basic primitives consist of the basic building blocks you can use to build almost anything. These include spheres, cones, cubes, and cylinders. The Extended subcategory holds more complex versions or compilations of the basic primitives.

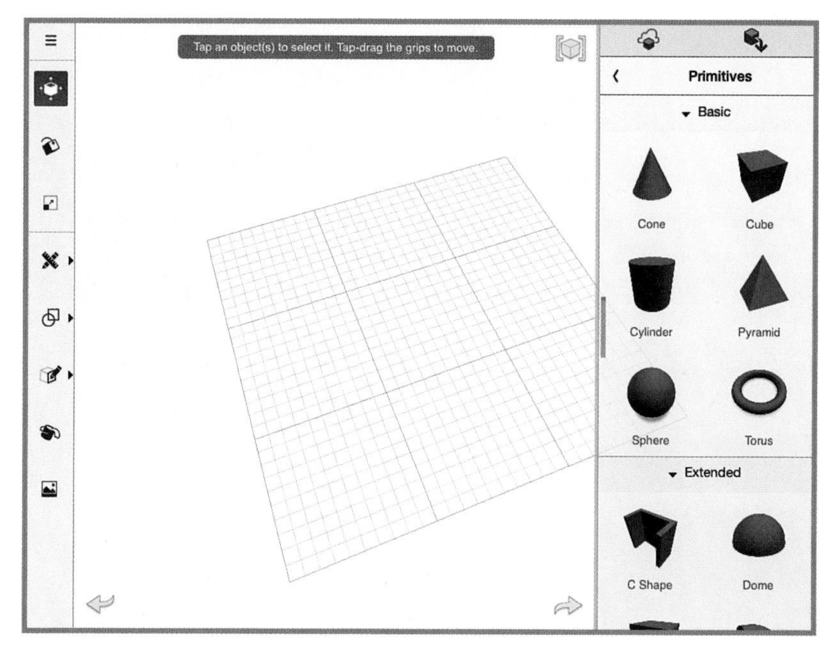

FIGURE 5.5 Basic and extended primitives.

You can swipe right-to-left on the border of the Primitives library panel to drag it out for full-screen viewing of the parts library (see Figure 5.6).

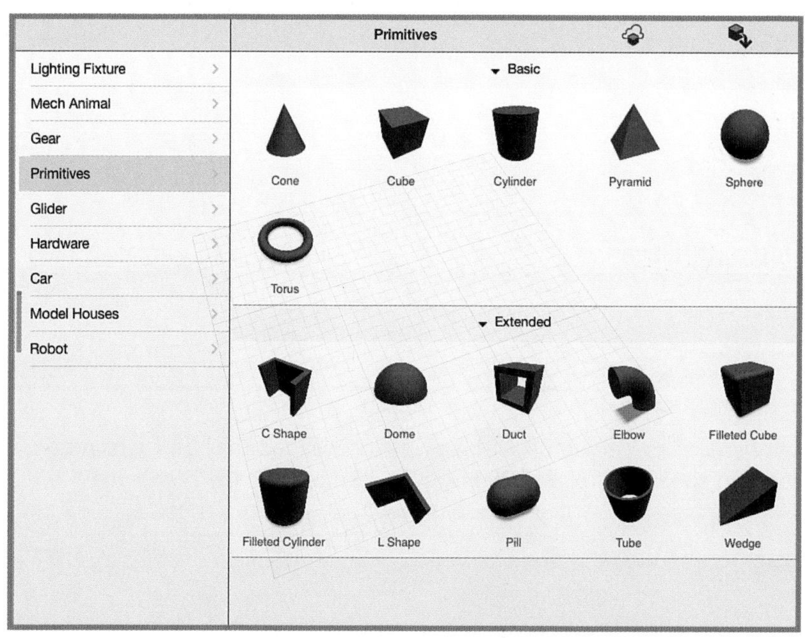

FIGURE 5.6 Full-screen viewing of the Primitives library.

The two icons at the upper right of the Library pane access the Parts Kit Library (via the Internet) and insert one of your existing project designs into the current one. Tap the icons a second time to return to the Primitives menu.

Parts Kit Library

The Parts Kit Library contains hundreds of additional parts and designs you can access with an Autodesk account (see Figure 5.7). You simply browse through the selection and tap on the part you want to add, and it begins to download. You then can position it on the workspace.

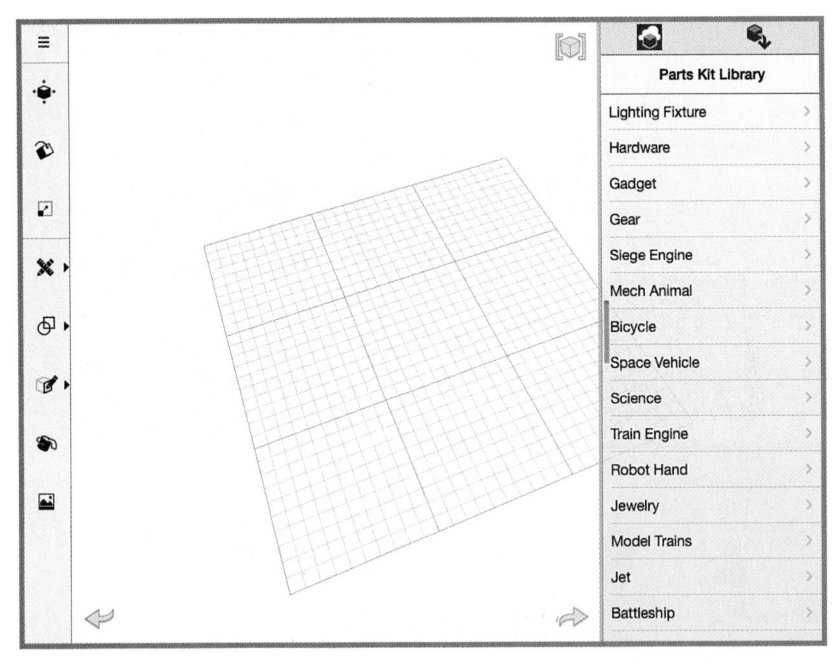

FIGURE 5.7 The Parts Kit Library.

> **TIP**
>
> Some objects might require a Premium membership to download them. Unfortunately, there is no indication on the model of whether it is free or part of the premium membership.

The left menu contains three primary object controls:

- Move
- Rotate
- Scale

These controls enable you to adjust your object with the arrows (called *grips*) that appear after the object has been selected by tapping on it (see Figure 5.8). You need to tap on an arrow and (without lifting) drag your finger in the chosen direction to move, scale, or rotate. Tap and then drag in the desired direction.

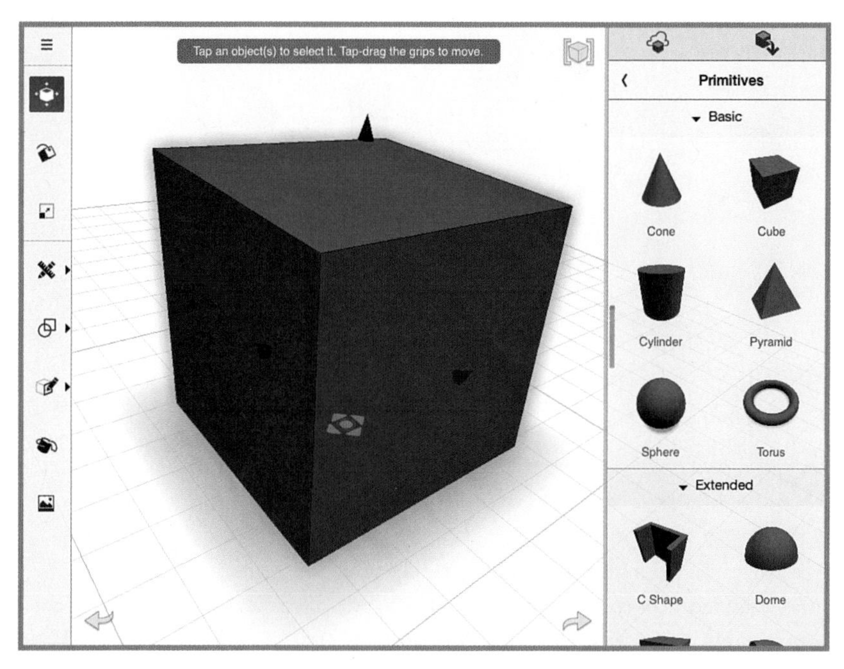

FIGURE 5.8 The Move tool and the activated grips.

TIP

The grip arrows might not always be showing both ways depending on the orientation of your object, but they work in both directions.

If you tap and drag your finger on a space not touching your object, you'll rotate the view or camera angle of the workspace. This is useful because you'll want to inspect your object from all angles. If you pinch and zoom your fingers, you move the virtual camera closer or farther away from your object and the workspace.

Rotate Tool

The rotate tool shown in Figure 5.9 enables you to rotate the object within the workspace. This is useful for positioning multiple parts together at the correct angle(s).

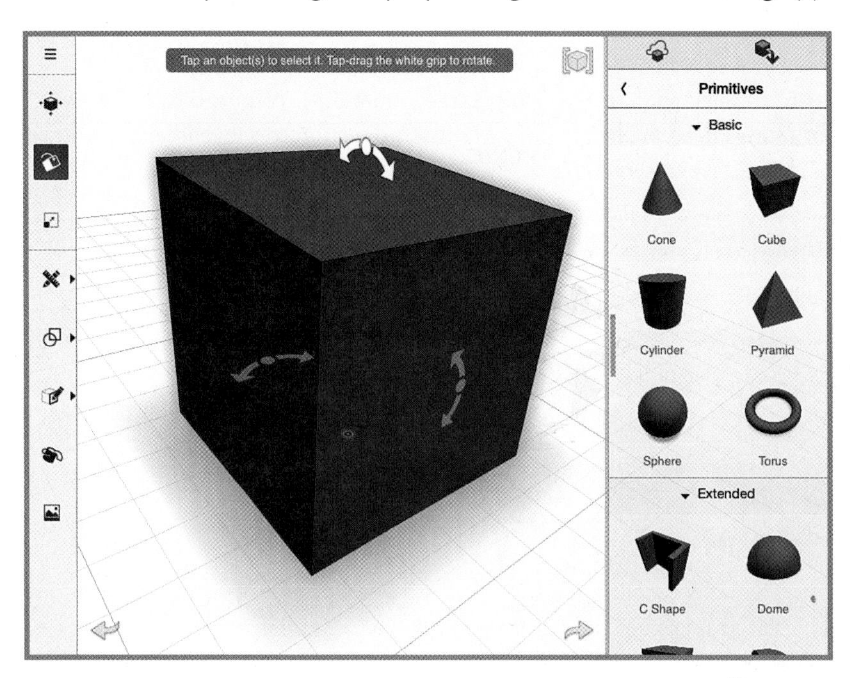

FIGURE 5.9 Use the rotate tool to alter your perspective of an object.

If you get zoomed or rotated into a view that doesn't help you or is confusing, you can always tap the Home View icon on the upper right of the workspace (it might be just to the left of the Parts Library window if it's open). This effectively resets the camera's view and centers your object in the workspace.

Scale Tool

The Scale tool shown in Figure 5.10 enables you to resize your object larger or smaller by tapping on it and then dragging the grips in the desired direction.

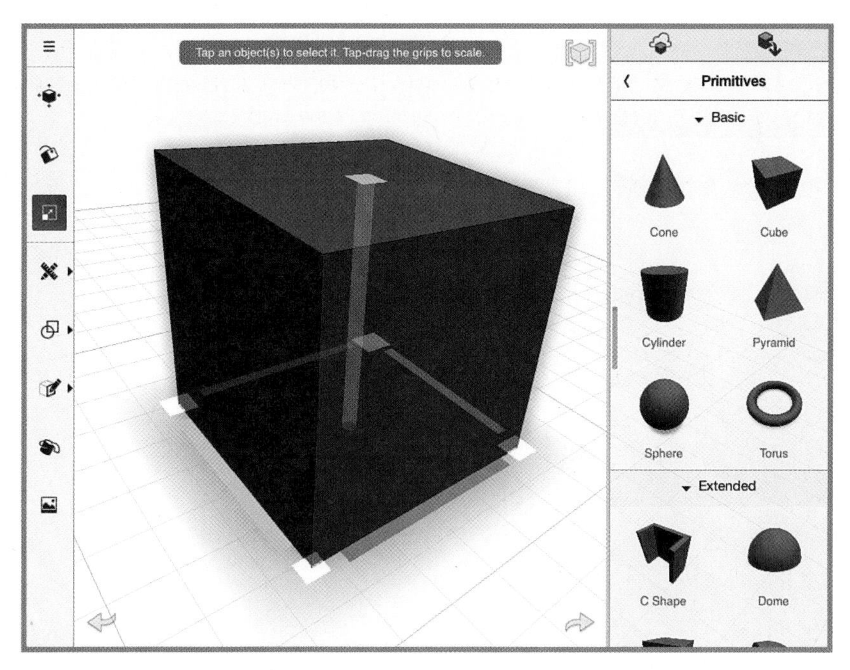

FIGURE 5.10 Scale tool.

Below the three primary controls are three additional controls that have submenus that enable you to adjust, combine, and reshape your objects.

Adjust

The Adjust submenu contains the Snap, Align, and Information tools. The icon for this menu changes depending on which submenu tool is selected.

The Snap tool lets you tap on one object and have it snap (move) into place on another object depending on where you tapped. This can help with the assembly of separate parts and make it easy to move objects around by simply tapping on the faces you want to position together (see Figure 5.11).

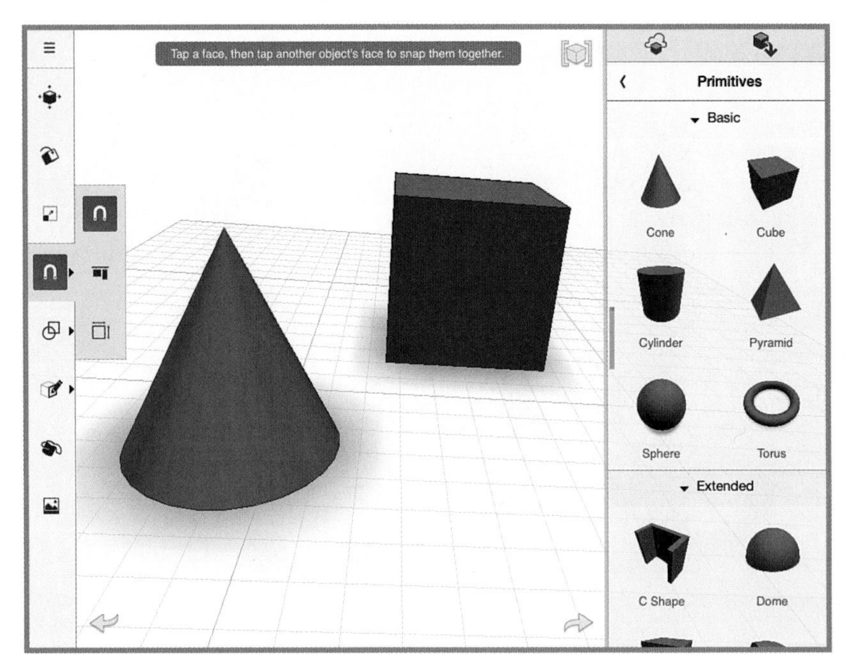

FIGURE 5.11 Snap tool.

You can use the Align tool shown in Figure 5.12 to assist in making multiple objects align along the same plane. When it's selected, the Align tool adds handles around your objects, and you can tap on the top, middle, or bottom handle for each plane of alignment. Doing so moves each object to line up with the others along the selected plane.

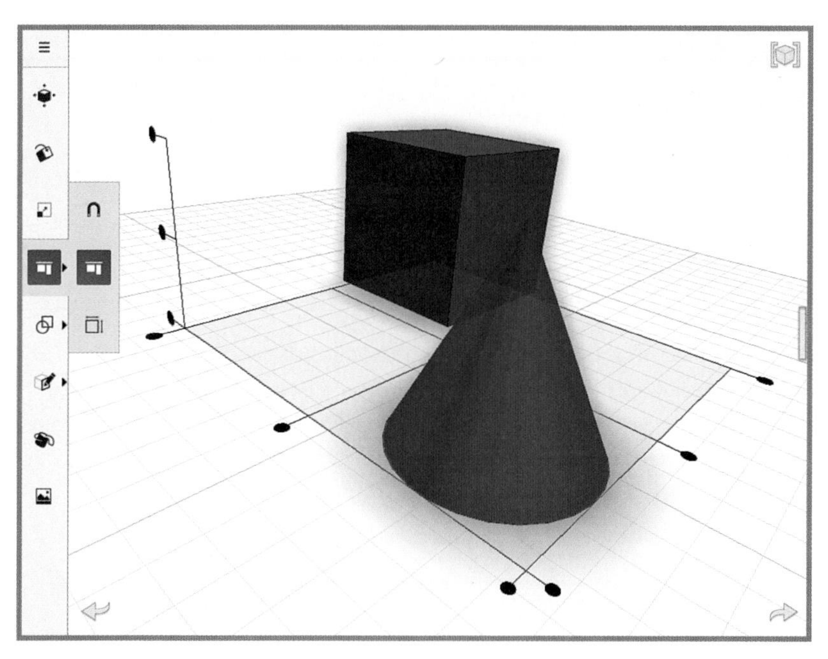

FIGURE 5.12 Align tool.

The Information button, which is the final Adjust menu option (see Figure 5.13), is useful to display a visible ruler showing you the dimensions of the selected object.

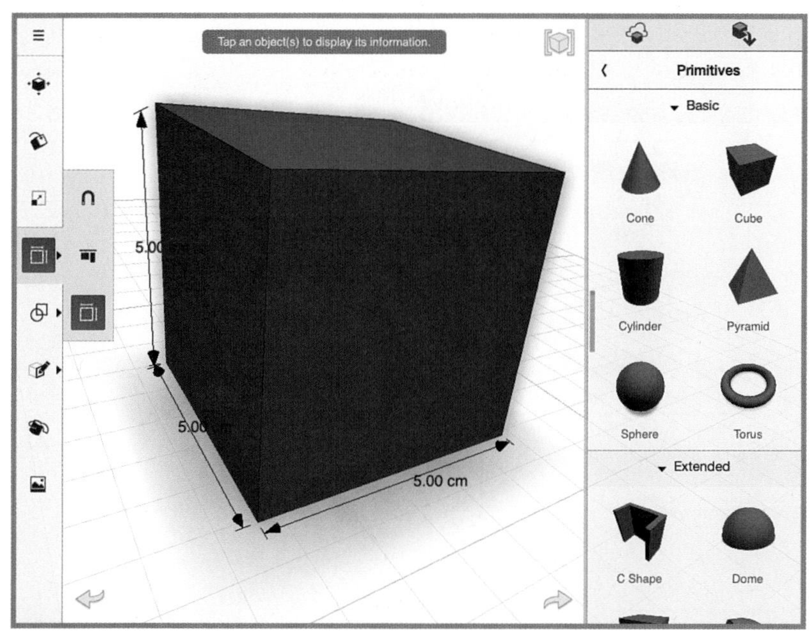

FIGURE 5.13 Tap to display an object's dimensions.

NOTE

Displaying the ruler and part dimensions is important if you're designing models that need to be used with existing parts where their size will matter when matching the parts together.

Combine Tools

Next we have the Combine tools. The icon for this menu changes depending on which submenu tool is selected. These controls enable you to combine two or more objects into one as well as subtract one or more objects from each other. These are what's referred to as *Boolean operations* and are discussed in later chapters.

If you position two objects together, you can combine them using the Combine button (see Figure 5.14). This merges the two objects into a unified, single object.

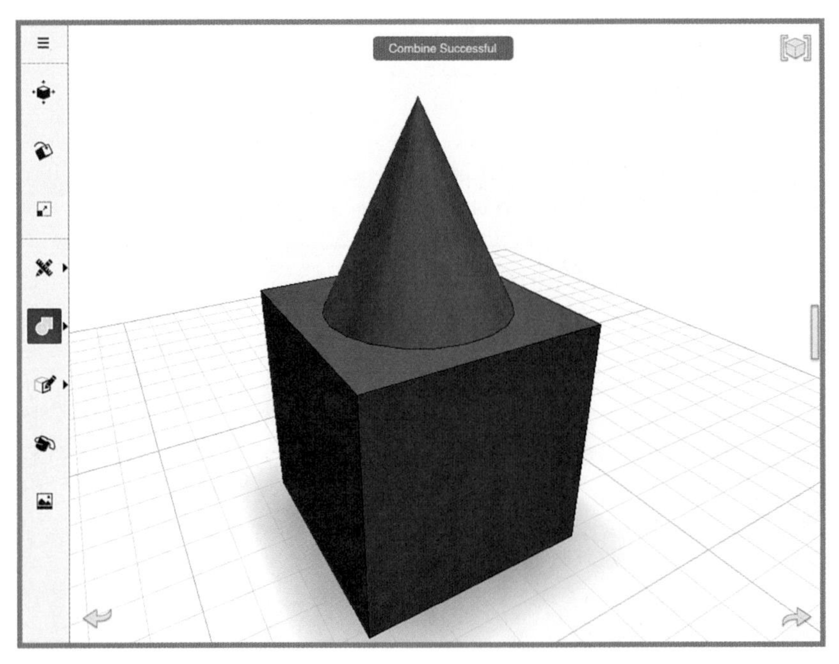

FIGURE 5.14 Select two objects to combine them into one.

Subtract Tool

Like the Combine tool, the Subtract tool requires two objects that are intersecting each other. By using the Subtract tool, it prompts you to select the first object; then the second object you choose is subtracted from the first.

In Figure 5.15, a cube has a cone positioned near the top of the cube but is below the surface of the cube.

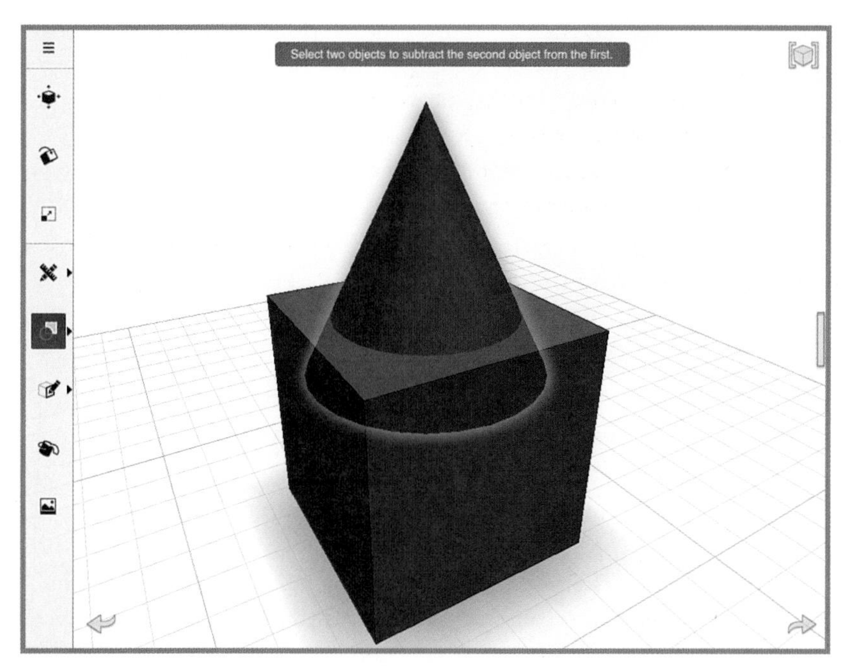

FIGURE 5.15 Selecting two objects with the subtract tool.

After subtracting, you can see the hole left behind where the cone was overlapping the cube (see Figure 5.16).

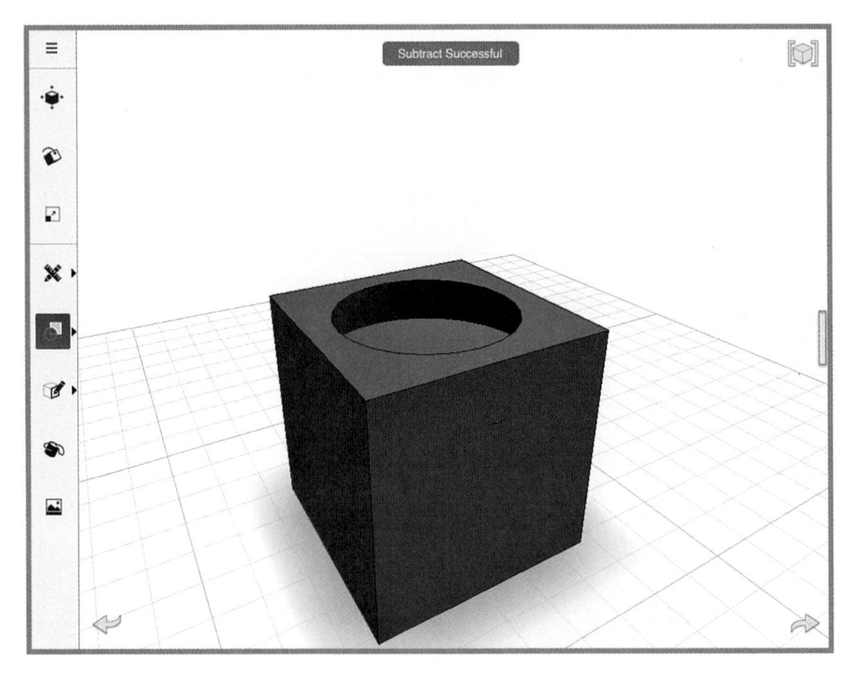

FIGURE 5.16 The cone was successfully subtracted from the cube.

Reshape Tools

The Reshape tools are next. The icon for this menu changes depending on which submenu tool is selected. These controls enable you to modify the objects you've placed in a number of ways.

Chamfer (bevels edges) and Fillet (rounds corners) work similarly in that you select an edge of your object and can choose to apply a rounded fillet (see Figure 5.17) or a hard-edge chamfer (see Figure 5.18) to your object.

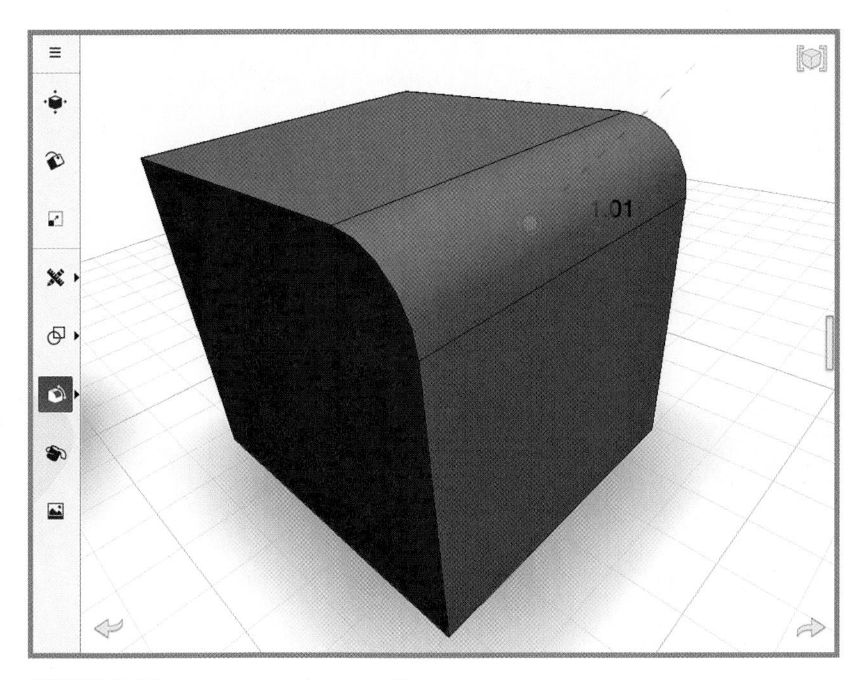

FIGURE 5.17 Tap an edge to fillet it.

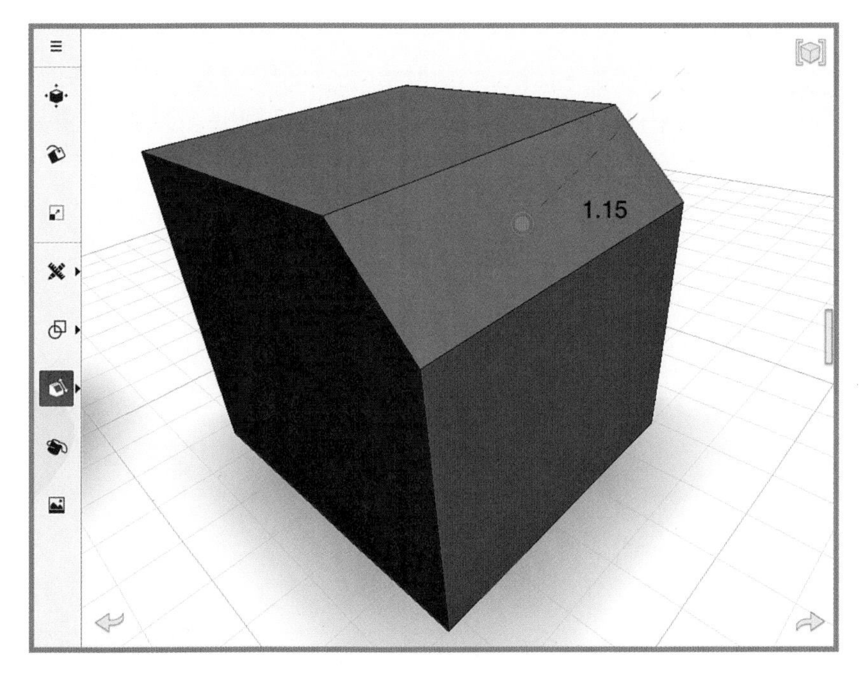

FIGURE 5.18 Chamfer.

The Push In/Out tool allows you to select a face on an object and pull it out or push it in. This makes for easy resizing of a particular face. but only on an object. In Figure 5.19, a face of the cube has been selected and you are able to tap/drag the face away from the cube to lengthen or shorten that face.

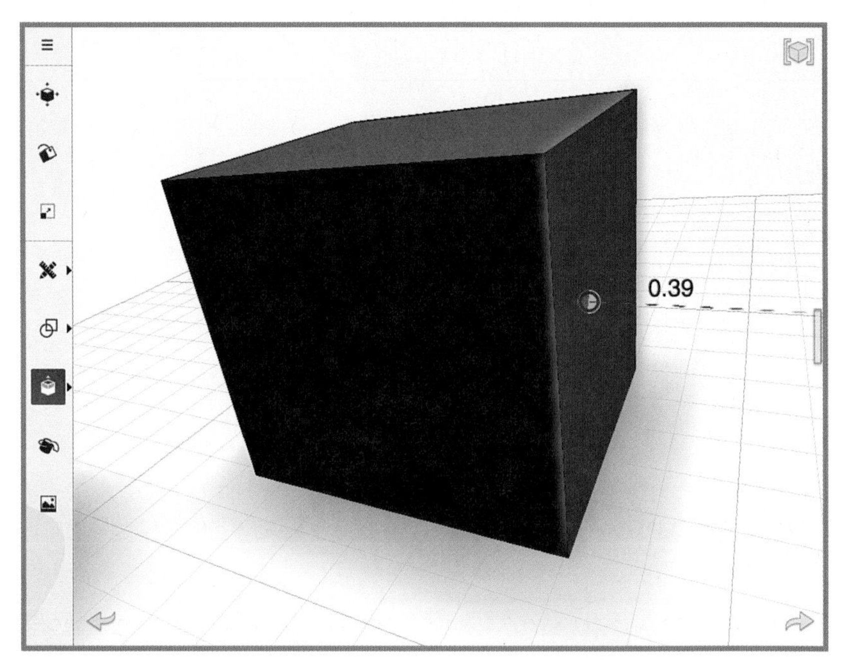

FIGURE 5.19 Tapping on a face enables you to pull it out or push it in to resize it.

Similarly, the hollow or shell tool allows you to select a face but instead of changing its length, it lets you create a hollow space with a varying thickness depending on your drag motion (see Figure 5.20).

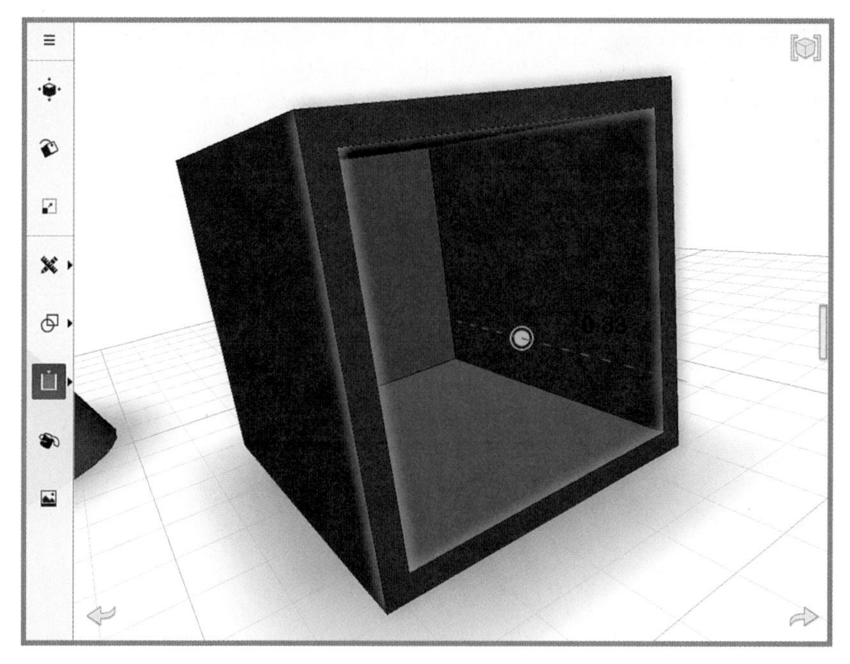

FIGURE 5.20 Hollow an object by dragging the toggle to adjust the size of the void created by varying the thickness of the shell.

The color change tool is useful to make more complex shapes more visable, especially if there are multiple parts because you can make each object a different color (see Figure 5.21).

The last two options on the left menu are the Change Color and Take a Picture controls.

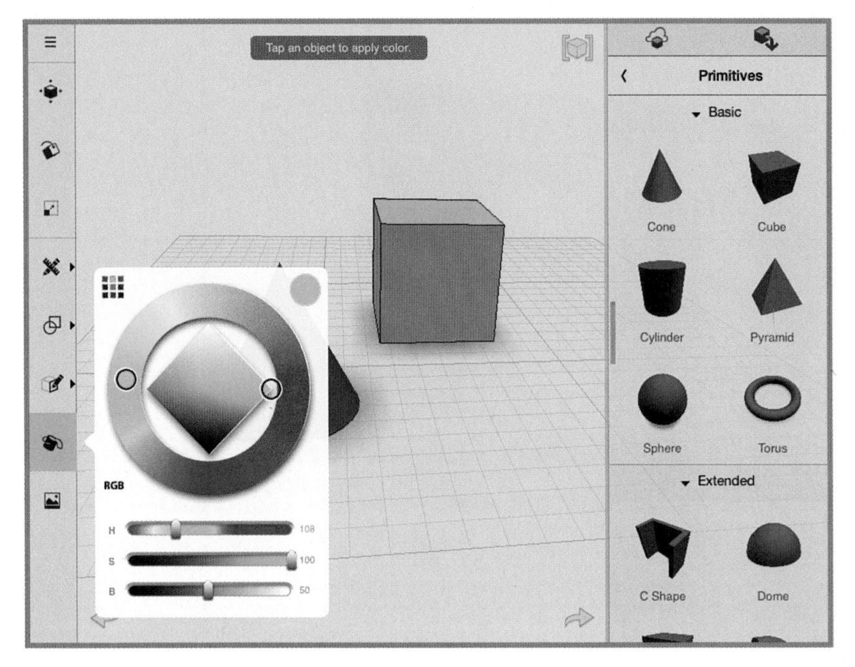

FIGURE 5.21 Changing the object's color.

Take a Picture Option

The Take a Picture option enables you to remove your model from the workspace and lets you take a screenshot without the controls and other application elements in the shot. It's a good way to show off your object when you're done and to give others an idea of what the final object is supposed to look like when physically 3D printed.

Object Editing Tools

Tapping on an object and holding the tap for a second pops open the long hold menu (see Figure 5.22). This lets you lock the position and duplicate or delete the object you've tapped on.

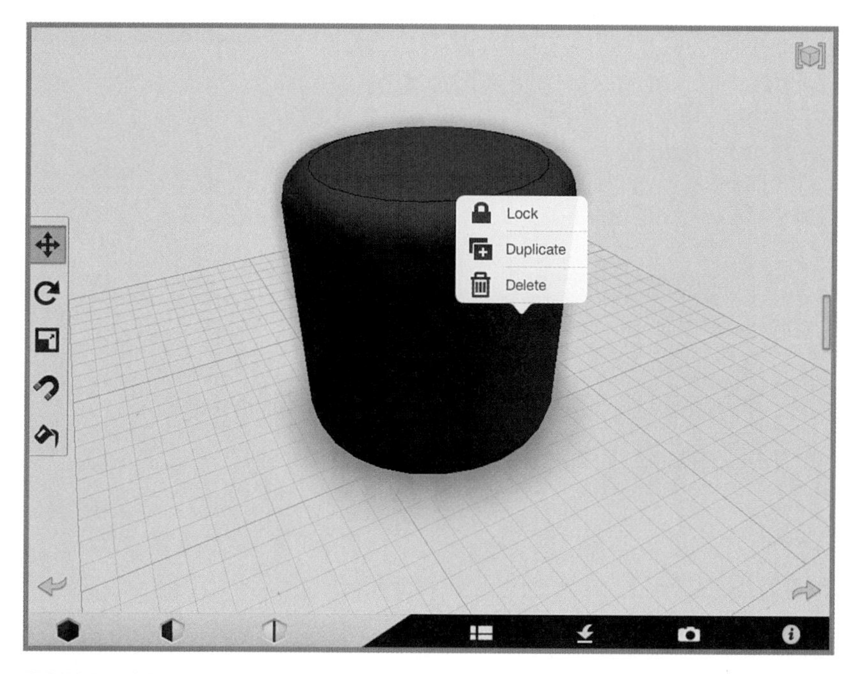

FIGURE 5.22 Long hold menu.

If you tap on the white arrow grip, you are presented with a pop-up numerical input (similar to a calculator) that enables you to enter a number for the amount you want to move or adjust your object without having to try to eyeball it using the touch interface. This allows you to properly position objects close to one another when knowing exactly how much your adjusting the object is important, as shown in Figure 5.23.

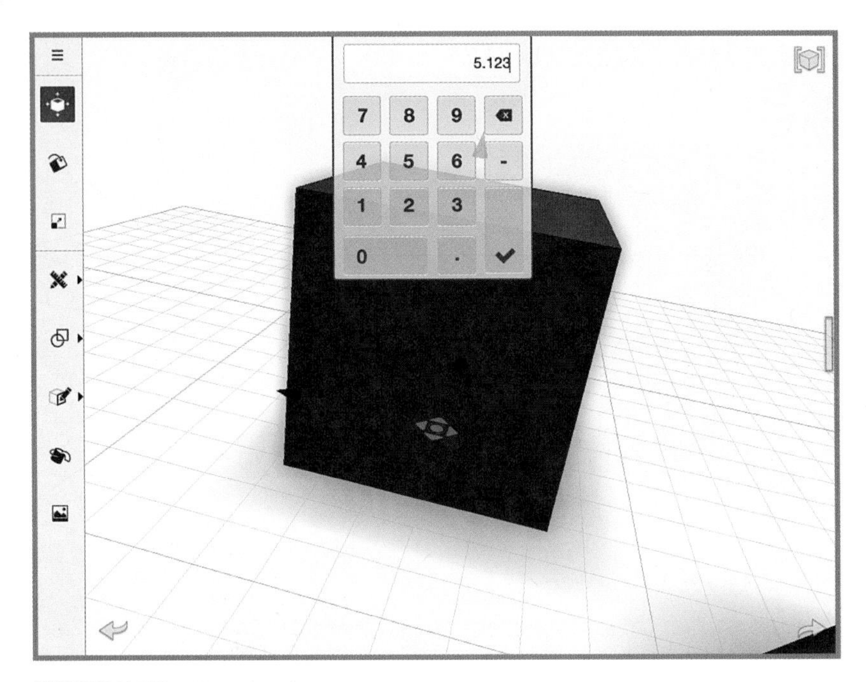

FIGURE 5.23 Precise input.

Project/File Menu Options

The Project/File menu shown in Figure 5.24 gives you a number of options to save your work and start a new project. The buttons for these options include the following:

- **New**—Start a new project. If you already have something on the go, it prompts you to save or discard your design before taking to you a blank workspace again.
- **Save to Cloud**—Saves your project in the cloud.
- **Save to this iPad**—Saves your project locally on the iPad.
- **Save a copy**—Saves a copy of your project locally on the iPad.
- **Create a 2D layout**—Premium feature that creates a 2D layout for use with other Autodesk tools, which won't be covered here.

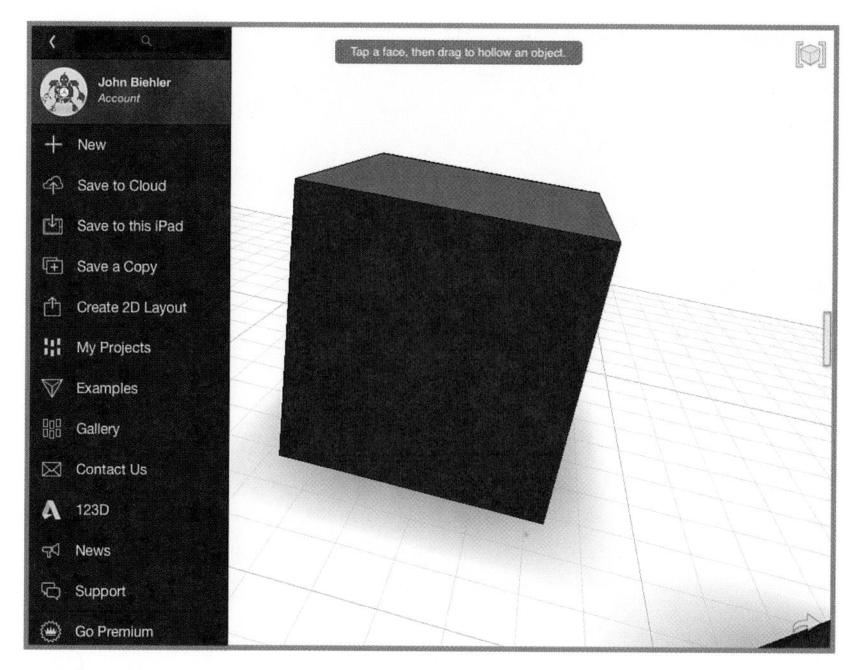

FIGURE 5.24 The menu allows you to save, export, and start a new project.

Projects and Galleries

The Projects and Galleries section of the menu (refer to Figure 5.24) give you access to the following options:

- **My Projects**—Displays all your currently saved projects on the iPad and in the cloud
- **Examples**—Displays all the models that Autodesk provides as examples, which can be great starting points to your projects
- **Gallery**—Displays all the publically shared projects in three different layouts (popular, featured, and recent)

Additional Support

The next section of this menu provides access to a variety of miscellaneous help and support options, including the following:

- **Contact Us**—Feedback email address for Autodesk
- **123D**—Displays other Autodesk products in the 123D suite
- **News**—Provides the latest news from Autodesk 123D
- **Support**—Takes you to a community support forum via your iPad's web browser
- **Go Premium**—Displays your current account information and details about upgrading to Premium if you haven't already

- **Help**—Displays simple onscreen help within 123D Design
- **About**—Displays legal and copyright information about the software and allows you to opt out of the collection of anonymous usage information for the application

TIP

The About section can be used to turn off Autodesk's ability to collect information about your usage of the software.

Saving to the Cloud

If you choose to save your object to the cloud (Figure 5.25), you'll be given options to share it publically, as well as to give it a title, a description, and some tags that describe it to others when searching the public galleries. If you choose to not make it public, the tags and description become optional fields. Press the Upload button to save it to the cloud.

FIGURE 5.25 Saving to the cloud.

Camera View

The Camera icon enables you to position your object onscreen without the workspace grid behind it so that you can share a screenshot via the iPad's sharing menu (see Figure 5.26). This allows you to easily email, message, or tweet an image of your project from right inside 123D Design.

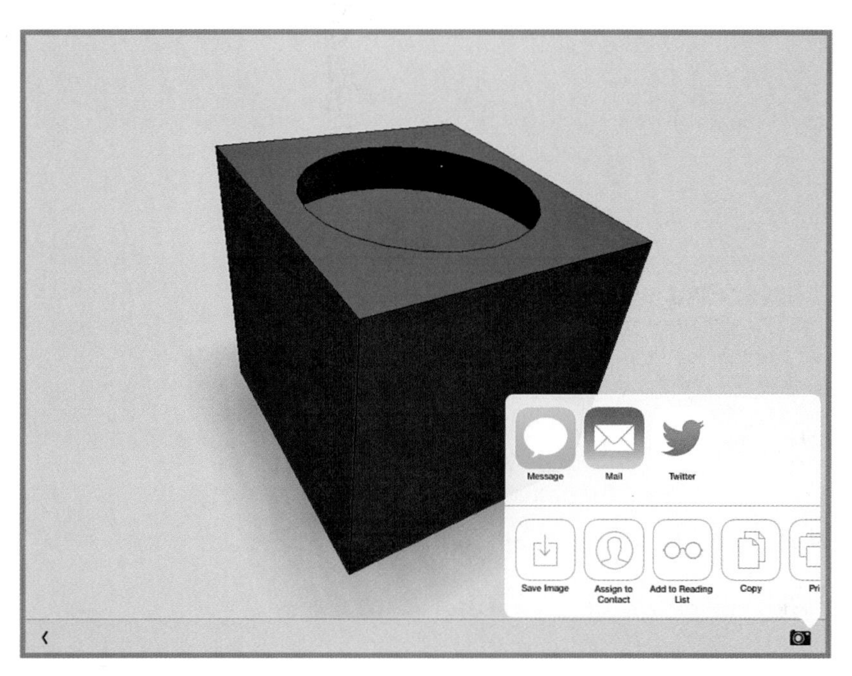

FIGURE 5.26 Camera view.

Summary

That wraps up our tour of 123D Design for iPad. In this chapter we've shown you most of the interface and control options you can use to create a model from basic primitive shapes. The Parts Library was also shown for starting with premade components that can be used as a starting point for your project.

In the next chapter, we walk through the steps to make a few different projects using what you've learned from this chapter.

123D Design Exercises for iPad

One of the most daunting things new users experience with design tools is the feeling of being overwhelmed with tools and options with no clear direction as to where to start.

In this chapter, we walk through the steps to create some simple objects that should help get you familiar with the purpose of the tools available to you.

Although the exercises are for creating simple objects, they should get you comfortable with using 123D Design and spark some ideas on how to create your own models for 3D printing.

Creating New Projects

The first project is to create a simple bowl. This bowl will also be the starting point for later exercises:

1. Create a new project and then slide open the Primitives panel on the right (see Figure 6.1). Tap on the cylinder shape (it looks like a soup can), which will drop it in the center of the workspace (see Figure 6.2).

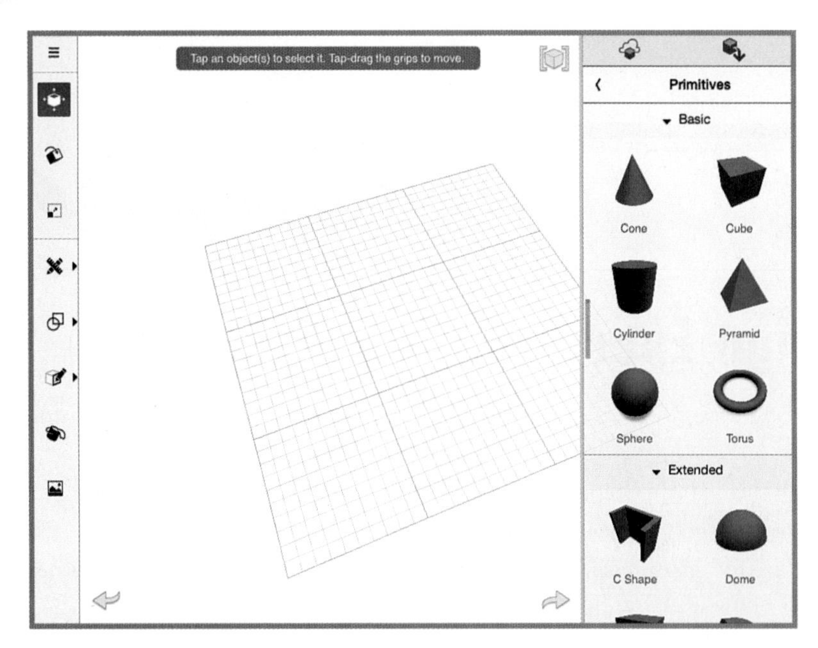

FIGURE 6.1 Starting with the Primitives menu.

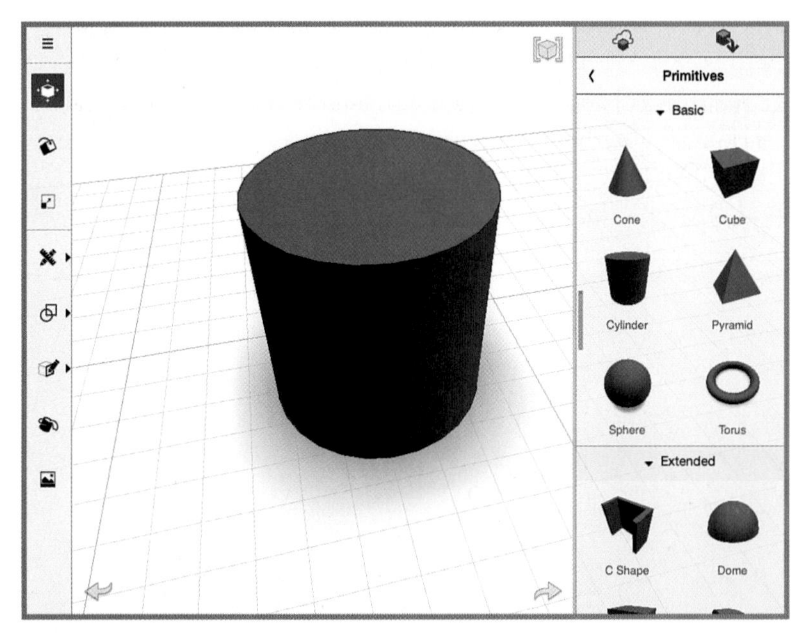

FIGURE 6.2 Cylinder in the center of the workspace.

2. Close the Primitives window by dragging it to the right (see Figure 6.3).

FIGURE 6.3 Primitives window closed.

3. Because we started with a solid cylinder, we're going to use the **Shell** tool to hollow out the cylinder to create the void in the bowl we're making.
4. Next, tap on the **Reshape** (Face Editing) menu button (third from the bottom on the left); then tap on the **Shell** button (the bottom button).
5. Now tap on the top of the cylinder and you should see a white grip appear; you can hold and drag the grip down to hollow out the cylinder (see Figure 6.4).

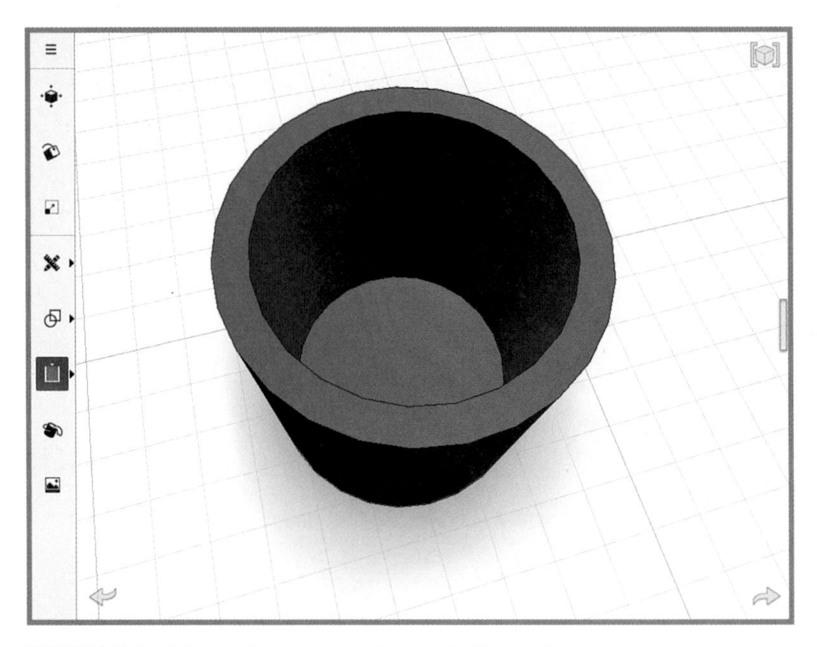

FIGURE 6.4 The cylinder has been hollowed.

6. Don't worry about how far down you go, but try to get as close as your can to the bottom. It won't let you go through the bottom (see Figure 6.5).

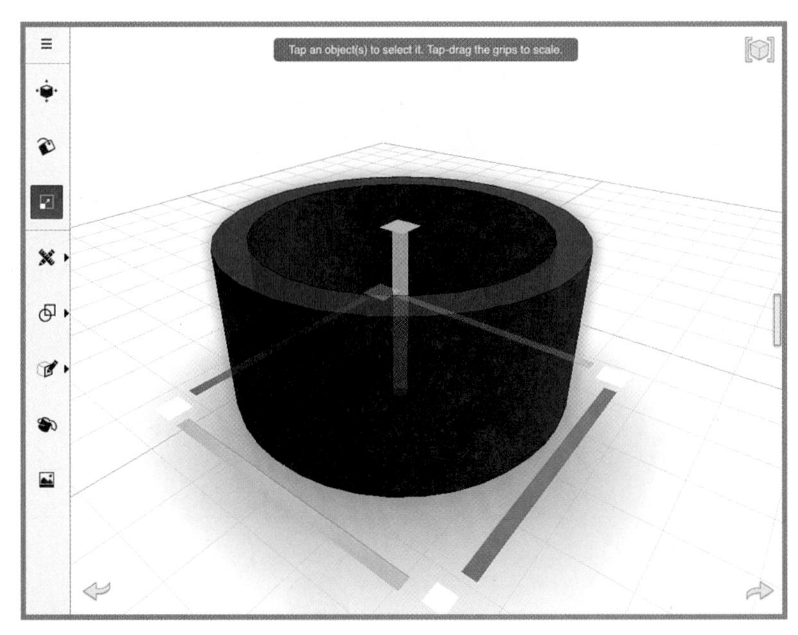

FIGURE 6.5 Scaling the hollowed cylinder.

Scaling and Smoothing Edges

Now that we have the basic shape started, we'll scale the cylinder to make it a little larger:

1. Tap on the **Scale** control on the left menu (fourth icon from top). White grips should appear; you can hold and drag them out or in to scale the cylinder.
2. Pull it out a little as shown in Figure 6.6. Keep in mind that it will scale the model in all directions, not just the direction you drag in.

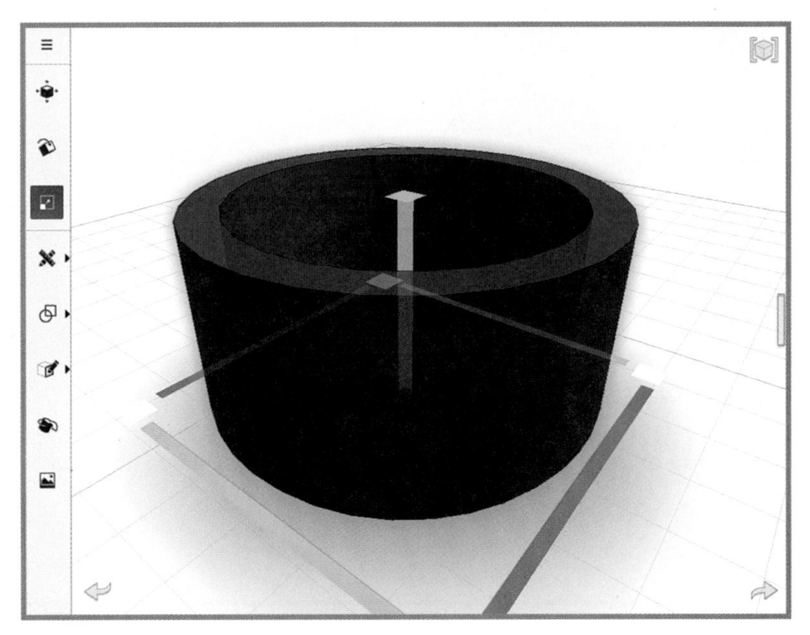

FIGURE 6.6 Scaled cylinder.

3. The next step is to smooth the bottom edges of the cylinder. Using the **Chamfer** tool, which is located in the Reshape (Edge Editing Tools) menu, tap on the bottom edge of the cylinder (see Figure 6.7). Like before, a white grip will appear and you can hold and drag it out or in.

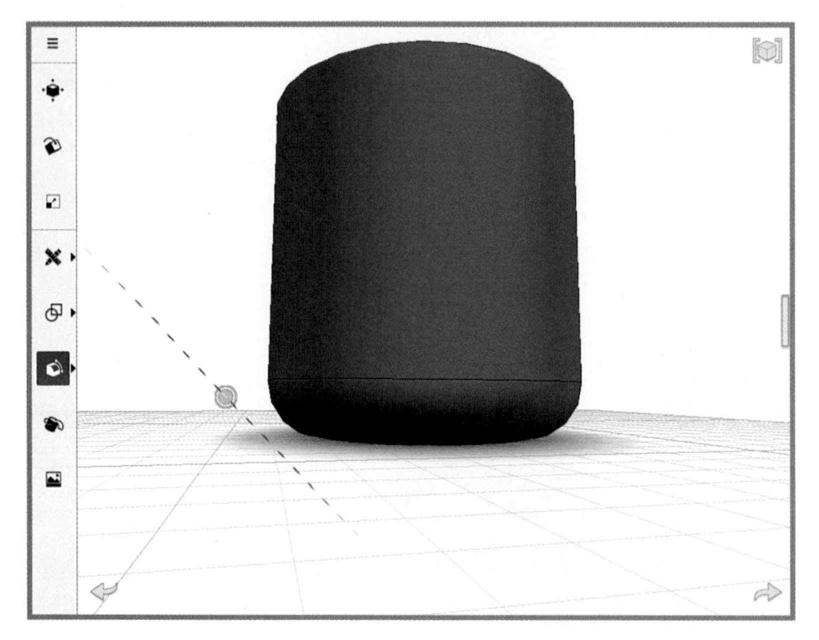

FIGURE 6.7 The cylinder now has a chamfered bottom edge.

4. You'll get an instant preview of the amount of chamfer applied. Use as much as you'd like your bowl to have.

TIP

Try to leave some part of the bottom flat so that it will sit flat on whatever surface you place the printed version on. If it's too chamfered, it might wobble easily or not even sit upright when in use.

5. Next, use the **Press/Pull** tool (second from the bottom in the **Reshape** menu) to pull down the top of the cylinder to shorten the height of your bowl (see Figure 6.8).

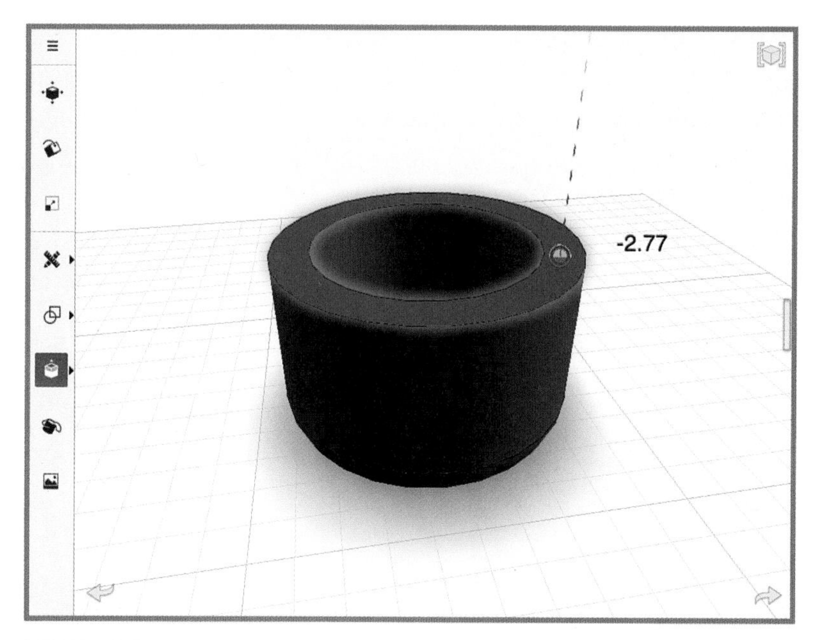

FIGURE 6.8 Press/Pull Tool.

6. After selecting the Press/Pull tool, tap and hold on the top of the cylinder and drag it down. Make it as low as you'd like, but try for something like what's shown in Figure 6.9.

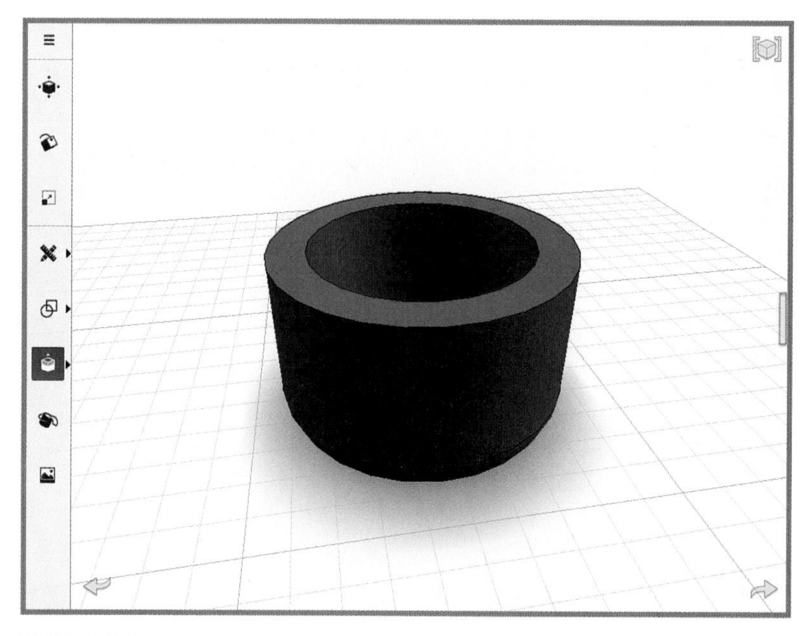

FIGURE 6.9 The shortened cylinder.

Using the Chamfer Tool

Now you could stop here and have a nice, simple model of a bowl. However, we'll continue using the **Chamfer** tool and smooth out the edges of the bowl:

7. Once again, select the **Chamfer** tool and then tap on the top face of the bowl. Drag the grip to smooth out the edges as desired (see Figure 6.10).

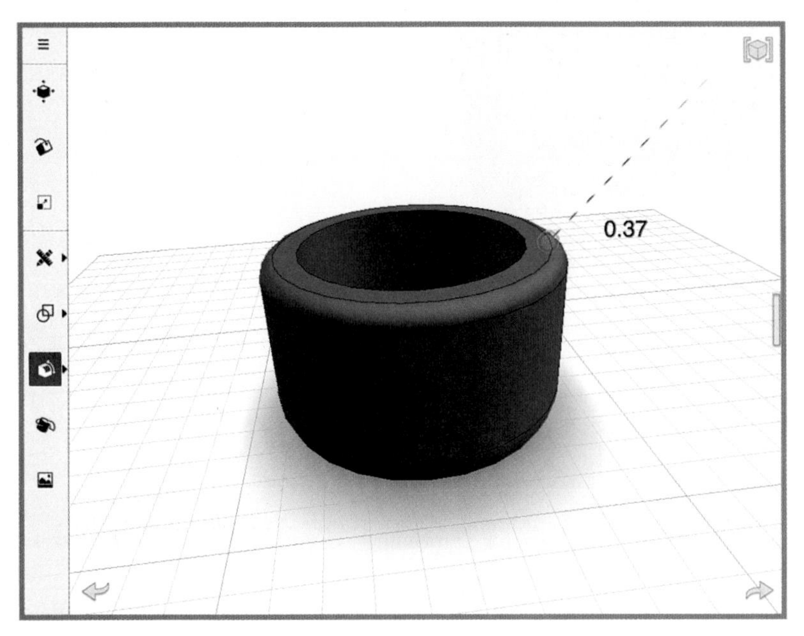

FIGURE 6.10 Chamfering the top of the bowl.

NOTE

You'll see edge lines in your model, but that is more for you to be able to continue to work on those edges within 123D Design. In reality, your model will likely be very smooth and rounded where you chamfered those edges and look great when 3D printed (see Figure 6.11).

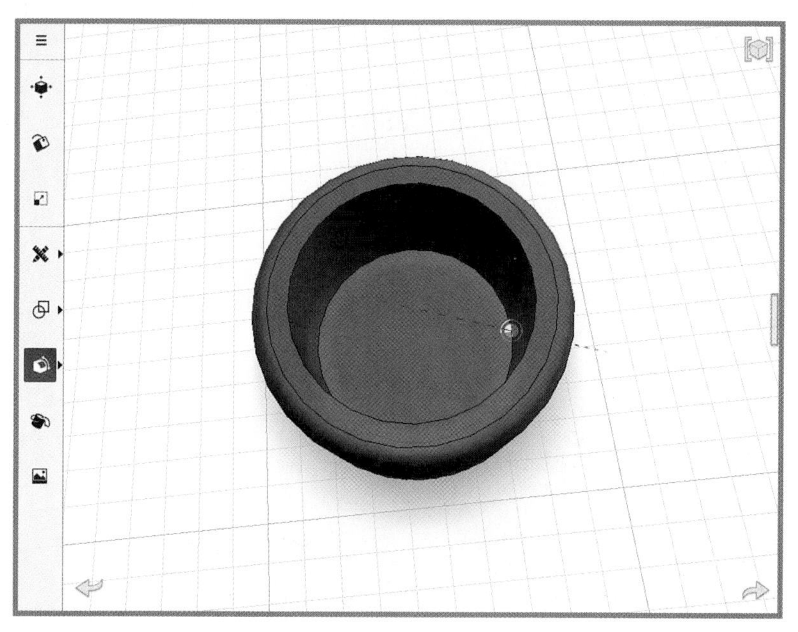

FIGURE 6.11 Chamfered top edge of bowl.

8. Inside the bowl, the bottom is still a 90° straight edge. Let's make it easier to get anything out of the bowl by making the bottom chamfered on the inside as well (see Figure 6.12). Save your project.

FIGURE 6.12 Smooth, chamferred inside bottom.

You should now have a nice smooth bowl, inside and out, that's perfect for storing change or small items.

To export the project for printing, you'll need to use the web or desktop version of 123D Design. We cover this in the next chapter. For now, save your project to the cloud, which will make it available to you in those other applications. Figure 6.13 shows the finished bowl when exported to STL and then loaded into MakerBot's MakerWare in preparation of 3D printing.

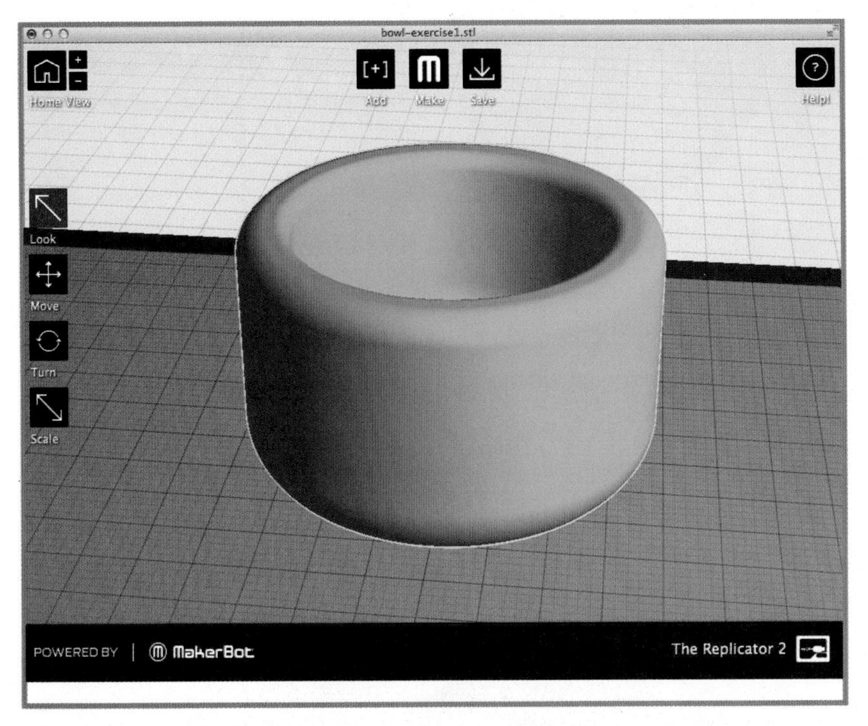

FIGURE 6.13 The finished bowl, ready to print.

Manipulating Existing Projects

Next we'll take the bowl shown in Figure 6.14 and turn it into a coffee mug. This will involve adding some parts to the bowl we just created and further changing the shape.

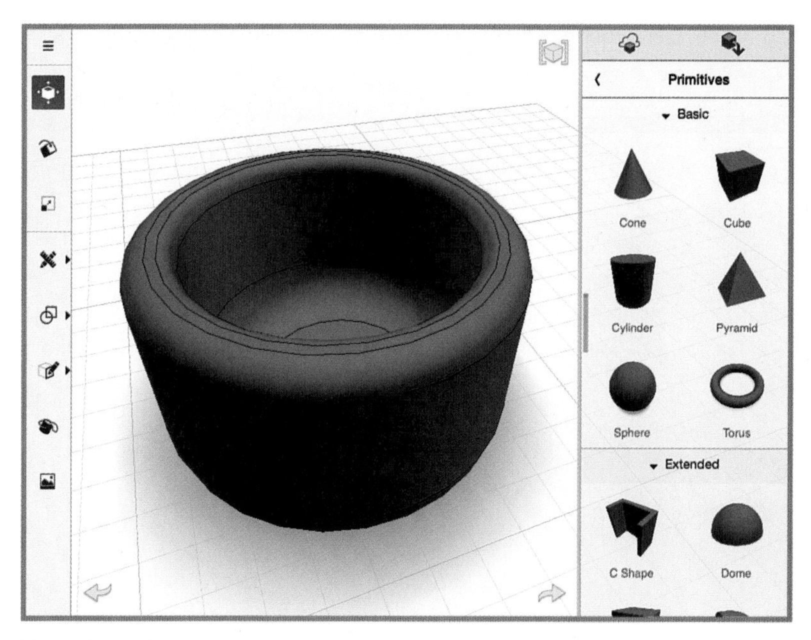

FIGURE 6.14 Starting with the bowl object.

1. Starting with your previously saved bowl model, drop a torus from the **Primitives** menu onto the workspace (see Figure 6.15).

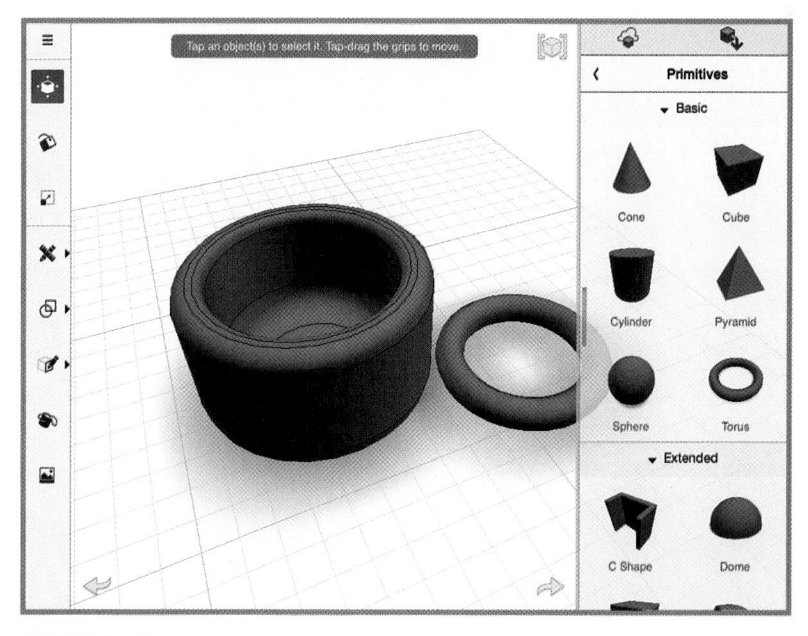

FIGURE 6.15 Adding a torus.

2. You can tap to drop it anywhere you'd like, and then close the Primitives menu by swiping to the right (see Figure 6.16).

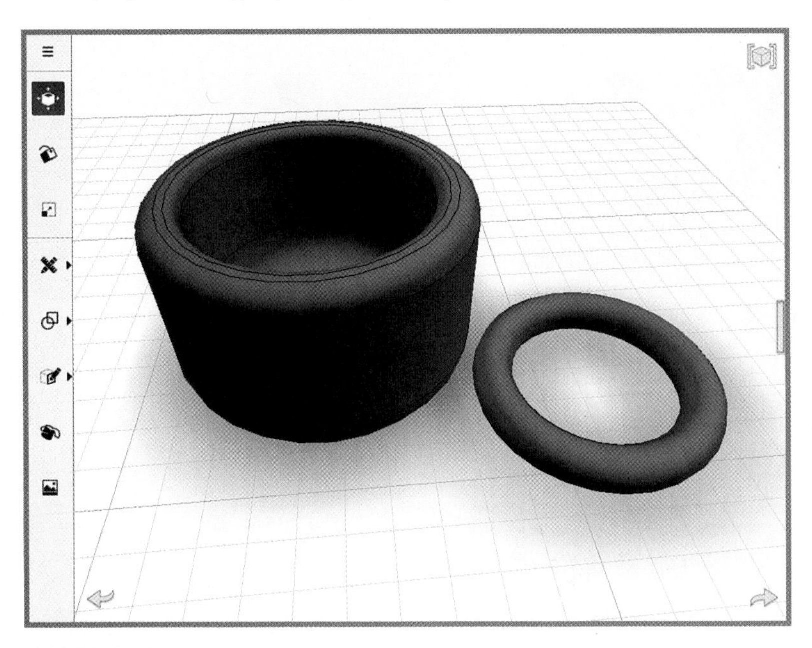

FIGURE 6.16 Bowl and torus.

3. We have to manipulate it a little before we attach it to the bowl model so you might want to move the torus a little away from the bowl to do that. You also might want to zoom out (see Figure 6.17) to be able to see all the controls and grips for what we're about to do.

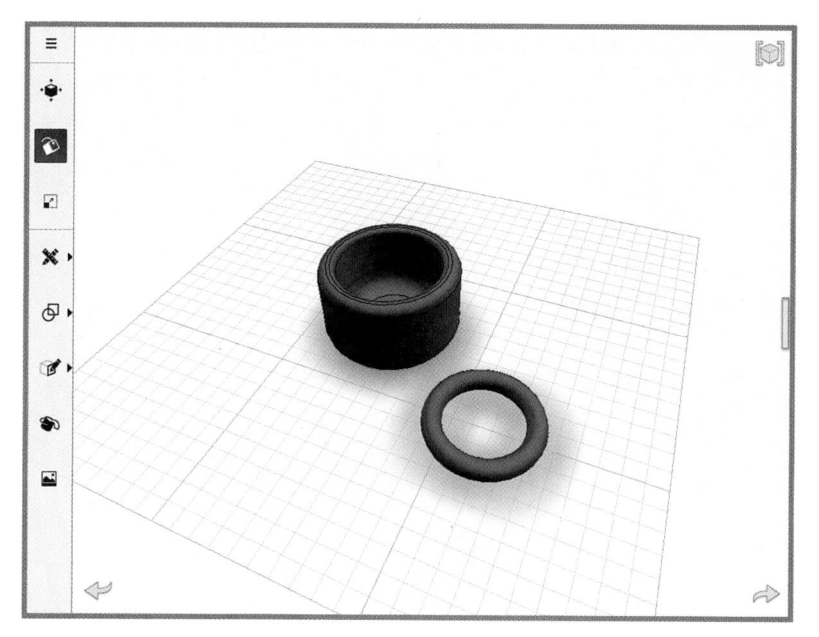

FIGURE 6.17 Zoom out to see everything.

4. The first change we'll make is to rotate it so that it is better aligned to be a handle for our coffee mug. Tap on the **Rotate** control on the left menu to display the white grip controls on the torus model.

TIP

Depending on your camera view, you might need to change your view to see all the rotate controls.

5. If your torus started out flat on the workplane, you'll need to rotate it 90° up from the workplane (see Figure 6.18).

TIP

You'll get visual feedback as to how many degrees of rotation you have as long as your camera view is zoomed out enough.

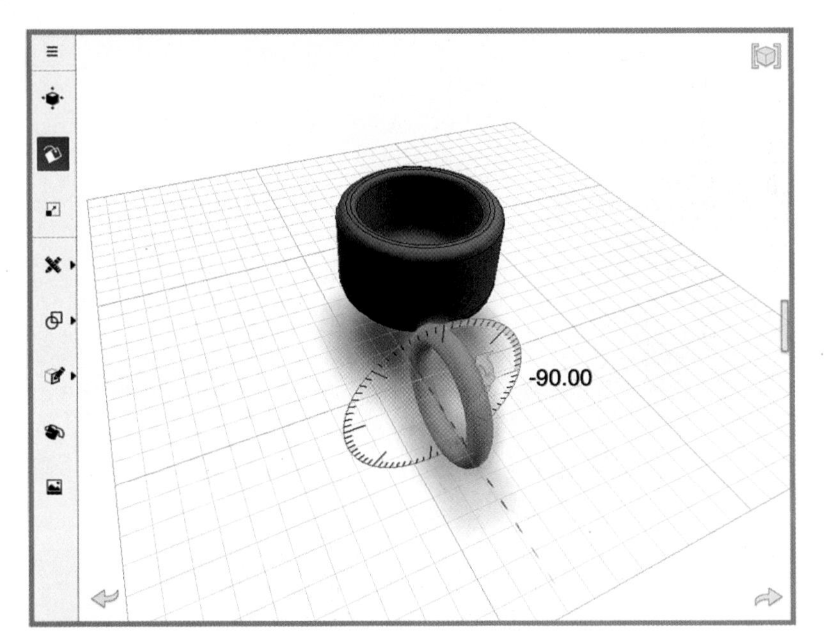

FIGURE 6.18 Rotating the torus.

6. You can flip between the rotate and the camera control if you need to adjust.

TIP

Keep in mind that if you do, any degrees of movement will be reset to the point you're currently at. It might be better to tap **Undo** if you're not sure whether it is perfectly 90° and start over.

Aligning the Model

When the torus is upright, you'll need to align it horizontally with the bowl model. In the case of Figure 6.19, the torus needed about 35° of rotation to line up with the bowl. However, this will vary depending on how you rotated your torus. Just try to align it with the bowl, as shown in Figure 6.19. Don't worry if the torus is partially inside the bowl model.

FIGURE 6.19 Aligning the torus with the bowl.

You'll likely need to change your camera view for this next move because we're going to move the torus up so that it is very close to being properly aligned with the bowl model:

1. Using the **Move** control from the left menu, tap the white up arrow, and drag it up.

As you can see in Figure 6.20, the torus is still a little large for a handle. We're going to scale the torus down to the right size using the Scale tool from the left side menu.

FIGURE 6.20 Oversized the torus.

2. Tap and drag the white grips that appear on the torus to resize it as desired (see Figure 6.21).

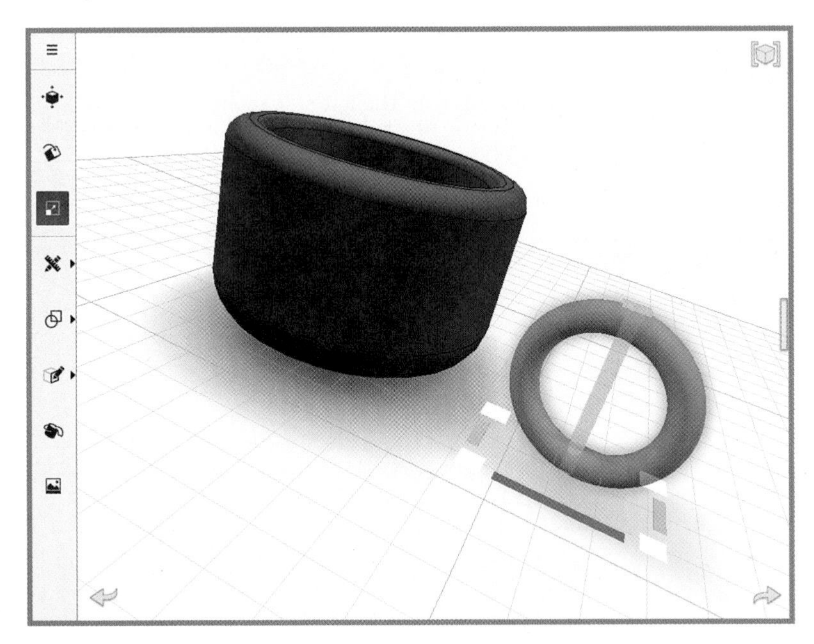

FIGURE 6.21 Scaled torus.

3. The scaling of the torus might have changed the vertical alignment to the bowl, so use the **Move** tool once again to position it as desired (see Figure 6.22).

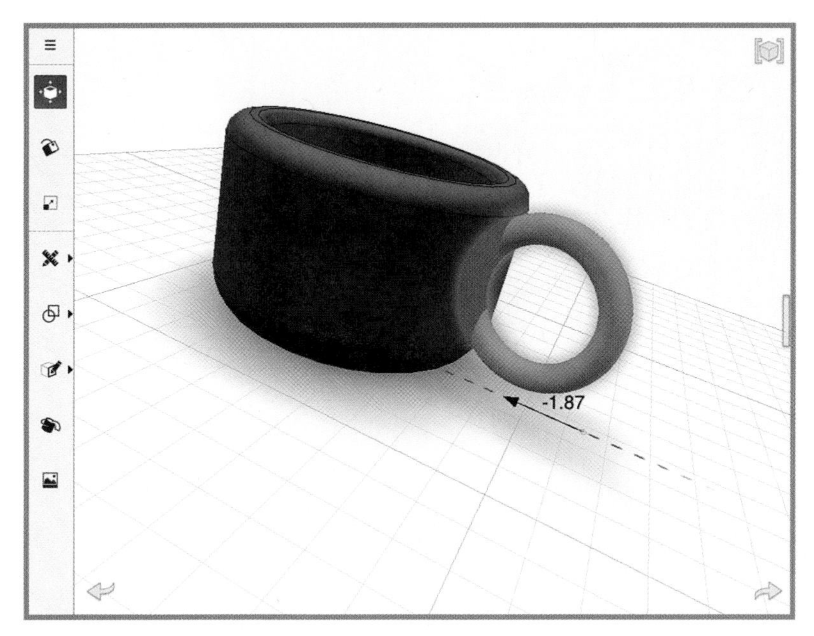

FIGURE 6.22 Moving the torus into position.

Finishing the Model

The last step to finish the model is to combine the torus with the bowl. Using the Combine tool from the Combine (Face Editing) menu, tap the torus so that it glows. This means you're going to combine the highlighted (or glowing) torus (see Figure 6.23) with the next object you tap.

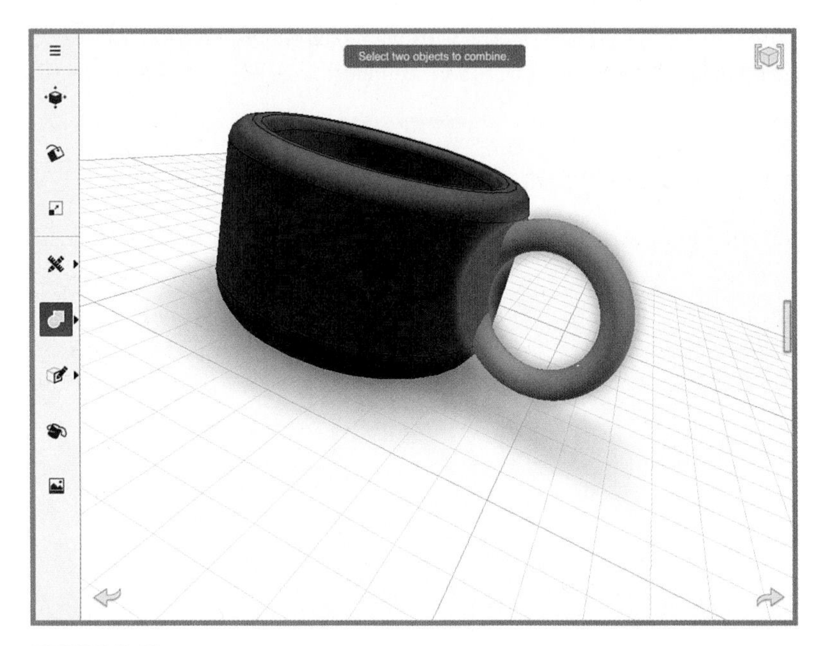

FIGURE 6.23 Highlighted torus for combining with the bowl.

Tap the bowl model and your combine should be successful (see Figure 6.24). The two objects have now been combined into one solid model.

FIGURE 6.24 Combine successful!

Congratulations! You now have a custom-made coffee mug...although unless you're printing it with a ceramic 3D printer, I probably wouldn't try to drink out of it. Don't forget to save your work.

Sending a File to a 3D Printer

Now you're ready to send the file to your 3D printer. There are a few things to consider with this particular model that will be useful to know when you're designing your own objects.

If you are using a plastic filament-based 3D printer (like a MakerBot Replicator), you will likely need to use support material if you try to print the coffee mug depicted here. Figure 6.25 shows how the supporting material is added in MakerWare for printing on a MakerBot Replicator 2 3D printer.

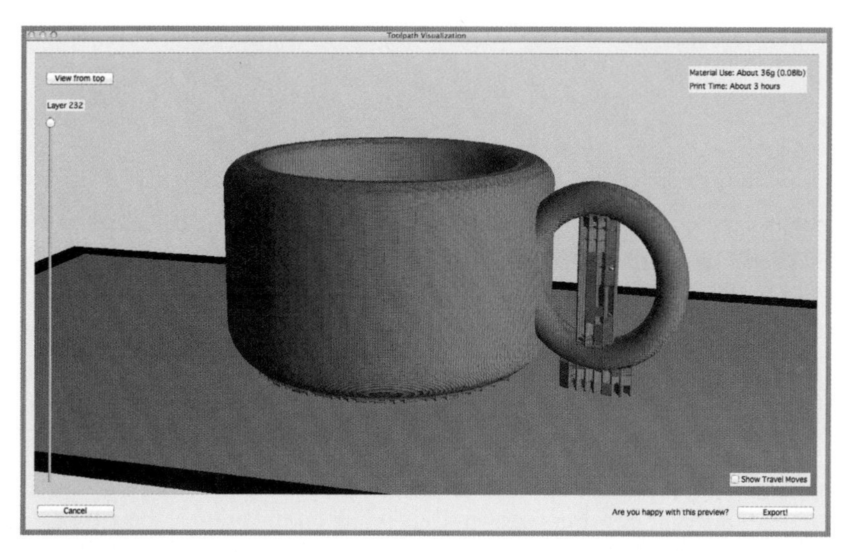

FIGURE 6.25 Support material for printing the coffee mug's handle.

Why is support material needed? This is due to the torus extending out from the bowl over empty space (see Figure 6.26).

FIGURE 6.26 Printed mug with support removed.

Generally, when 3D printing with an FDM-based printer, you need to ensure your angles are no greater than 45° from an edge so that the printer can build up material to print that area.

If there is no material from a previous layer to print upon, you will likely get plastic printing in air that might or might not droop down. Depending on the printer, this may be undesirable and cause the print to fail.

These limitations don't exist if you're using most other types of printers such as SLS, which has built-in support via the powder used to print with.

Trial and error with your specific printer will yield the best results.

Summary

In this chapter we created a bowl using a number of tools. We then continued to manipulate this model and turn it into a coffee mug by changing its shape and adding a handle.

In the next chapter, we move from the iPad to a desktop or laptop computer for even more options and controls with 123D Design.

7

Workspace Basics of 123D Design for Mac and PC

Continuing along the lineup of 123D software, we come to the PC and Mac versions of 123D Design (see Figure 7.1). These are the bigger brothers to the iPad sibling covered previously in Chapter 5, "Introducing 123D Design for iPad."

FIGURE 7.1 Like the iPad version, it's a free download from Autodesk.

→ *A free download of 123D Design for PC or Mac is available from Autodesk at http://www.123dapp.com/.*

In many ways it's similar to the iPad version. but it has an expanded range of features and options. Both Mac and PC versions connect to the Autodesk network and work best with an online account on which to store your designs.

For this chapter, we'll be using the Mac version of the software, but the controls are similar in the PC version.

123D Design Templates

When you first launch 123D Design, you'll be presented with a splash screen that contains essential tips (see Figure 7.2), quick start templates (see Figure 7.3), and an option to open a project from the Gallery (see Figure 7.4).

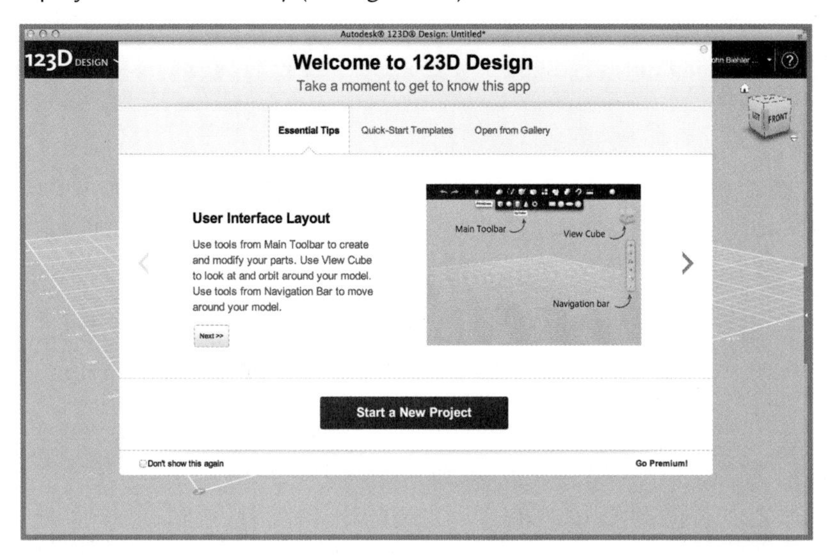

FIGURE 7.2 Welcome splash screen.

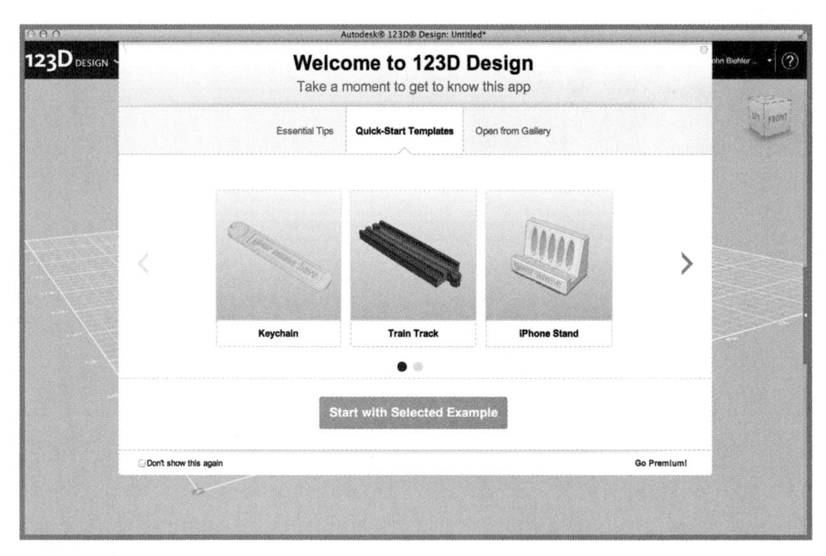

FIGURE 7.3 Quick start templates.

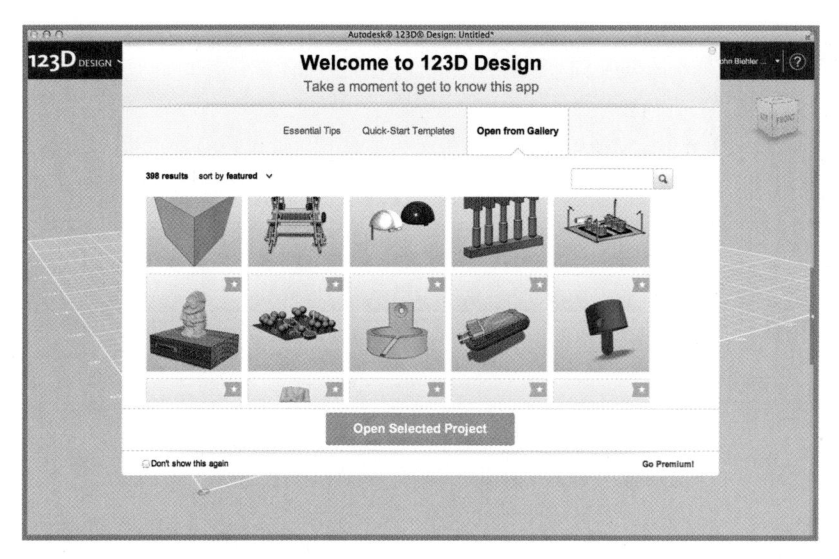

FIGURE 7.4 Gallery projects.

The essential tips give you an overview of the program's controls. The quick start templates are a collection of projects that Autodesk provides to give you some ideas of the types of objects you might want to make with the software. These templates can be used as the starting point for a new project because they are completely customizable.

As with the iPad version, the Gallery is full of publically shared projects that are available to download for free. Like the templates, these files are a great way to see how certain objects are constructed and serve as a perfect starting point for you to customize and make your own project from.

Starting a New Project

You can also start a new project by clicking the **Open a New Project** button at the bottom. If you prefer, you can prevent the splash screen from showing every time you start 123D Design by clicking the check box on the lower left of the splash window.

When you create a new project, you'll be presented with an empty workspace and a toolbar and menu along the top, as shown in Figure 7.5.

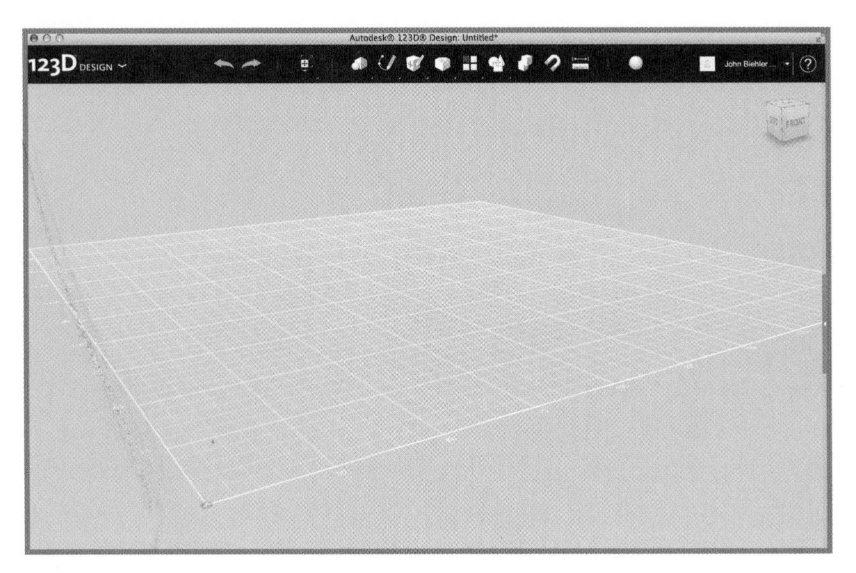

FIGURE 7.5 Blank canvas for a new project.

On the top left is a drop-down file menu under 123D Design (see Figure 7.6). This is where you can create a new project, open an existing one, insert an existing model into the workspace, save and export your project in various formats, as well as send your project to be 3D printed—either locally if you have a printer or to a third-party service.

FIGURE 7.6 Main menu.

The **Send To...** option enables you to send your project to other Autodesk applications including **123D Make** and the **CNC Utility**—both of which are not covered in this book. Essentially, they allow you to prepare your model for cutting using methods other than 3D printing.

→ *Detailed coverage of the other Send To... option of using a third-party 3D printing service is included in Chapter 16, "Using a Third-Party 3D Printing Service Bureau."*

Toolbar Controls

Let's go through the top toolbar controls now.

Starting immediately to the right of the file menu are the self-explanatory Undo and Redo controls.

Next are the **Transform** controls (see Figure 7.7). These include the **Move** and **Scale** options for the selected object in the workspace. The Move control allows you to move an object anywhere in the workspace as well as rotate it in any direction.

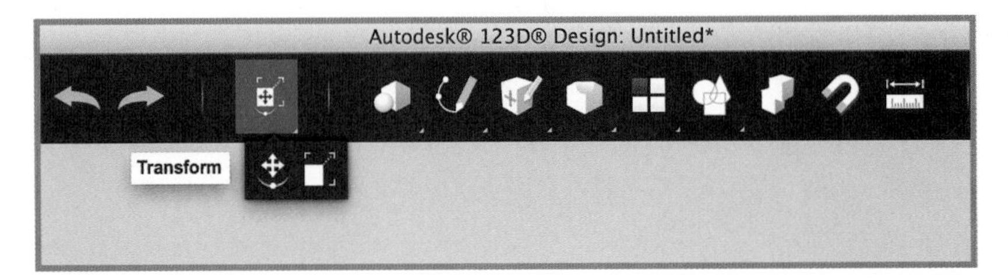

FIGURE 7.7 Transform.

When Move is activated, icons appear on the selected object (see Figure 7.8). Use your mouse to move the object by dragging the icons. You can also use your mouse to rotate the object manually in any direction, or you can use the pop-up degree indicator to specify a precise amount of rotation.

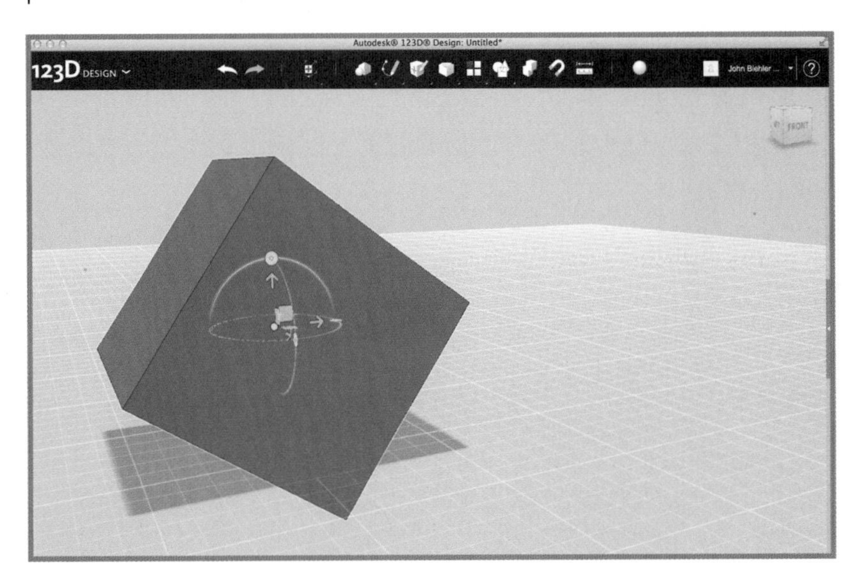

FIGURE 7.8 Rotating with the Move control activated.

In Figure 7.9, the Scale control is activated. It activates an arrow icon on the selected object, which you can drag to scale up or down. You can also type in a scale factor amount in the box that appears at the bottom.

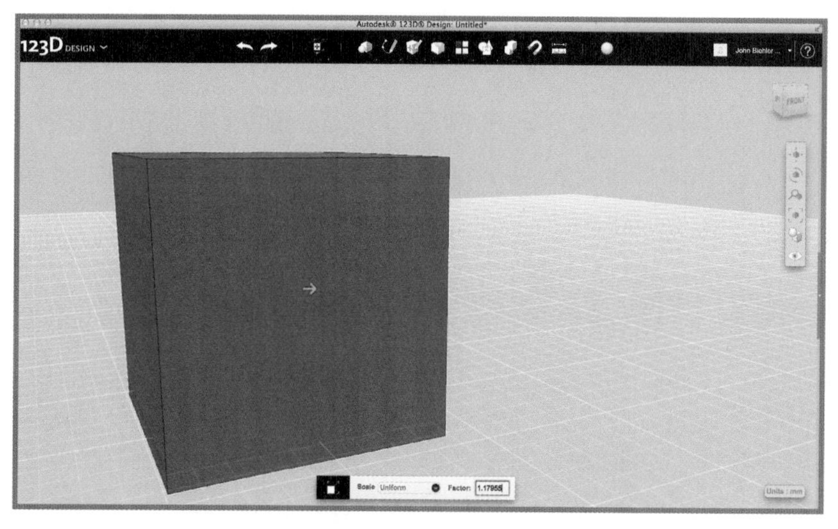

FIGURE 7.9 Scaling an object.

Menu Options

Next is the **Primitives** menu (see Figure 7.10). It contains a number of basic shapes with which you start creating your objects. Unlike the iPad version, however, this desktop version comes with more shapes.

FIGURE 7.10 Primitives menu.

After you drop a primitive onto the workplane, clicking it will activate a **Selection Based Options** menu near the object itself (see Figure 7.11). This is a submenu of controls and options for commonly used features, and it is copied from the main toolbar above.

TIP

The subheadings can vary depending on the selected object. For example, you might see different menu options when working with a 2D object than you do with a 3D object.

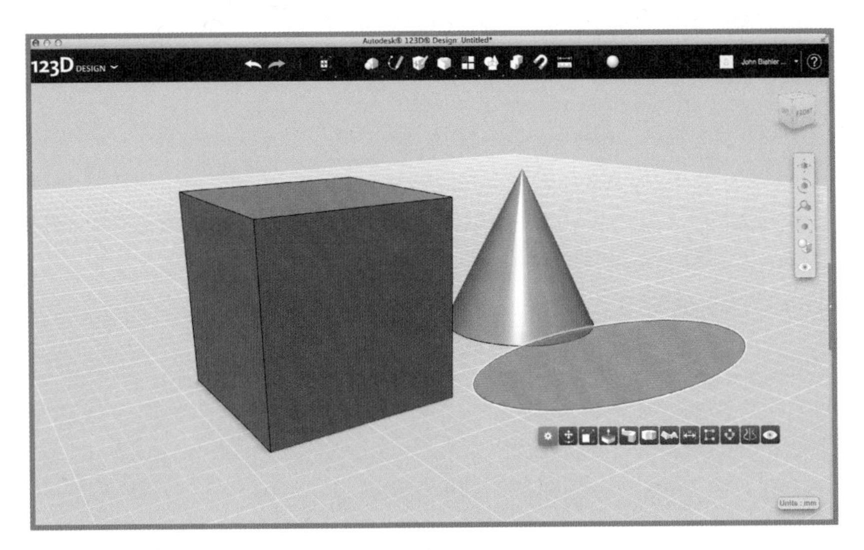

FIGURE 7.11 Selection-based options menu.

Next up is the **Sketch** menu and its controls (see Figure 7.12). These are the controls that enable you to draw various two-dimensional shapes (similar to the 2D primitives) that have no volume and then extrude them into 3D or use them as the framework for other objects. It's separated into three sections. The first section contains basic shapes followed by various line tools to modify those shapes using toggle points along the lines; finally some precision tools are shown for finishing the sketch object including trimming and extending.

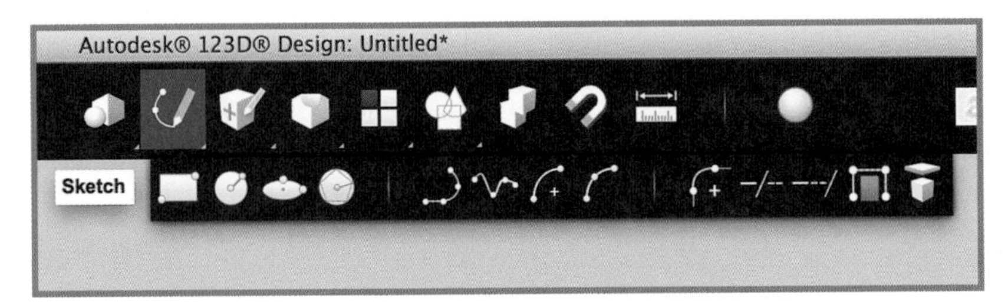

FIGURE 7.12 Sketch menu controls.

The next section is the **Construct** menu (see Figure 7.13). This is where you can take those primitive shapes and sketches to the next level using tools like **Extrude**, **Sweep**, **Revolve**, and **Flow**. These tools allow you to take an object or a sketch and pull it into different, 3D shapes along a specified path. We go over this more in the next chapter.

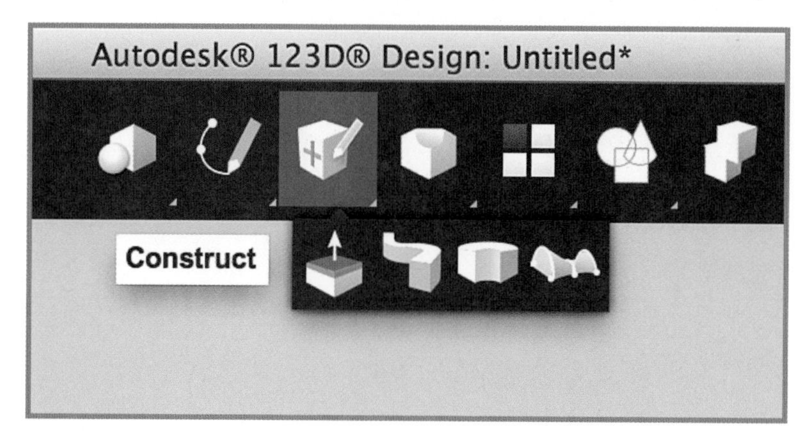

FIGURE 7.13 Construct menu.

Continuing along the menu bar, you'll see the **Modify** menu (see Figure 7.14). These controls enable you to fine-tune your objects in various ways, including smoothing out hard corners with Fillet or Chamfer and creating hollows inside the object using Shell.

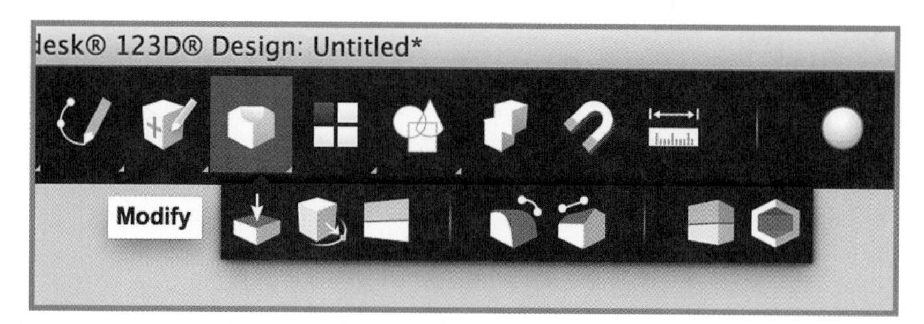

FIGURE 7.14 Modify menu.

The **Pattern** menu allows you to arrange your objects along a path and distribute them evenly (see Figure 7.15).

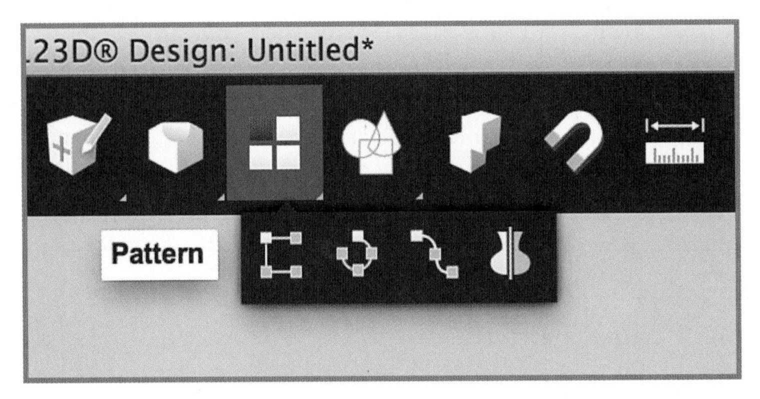

FIGURE 7.15 Patterns menu.

The **Grouping** menu enables you to organize your objects and lock them together to move them as a group (see Figure 7.16).

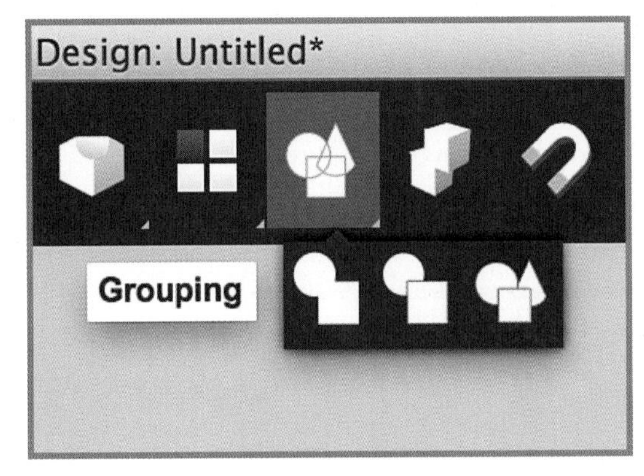

FIGURE 7.16 Grouping menu.

The **Combine** merges multiple objects into one (see Figure 7.17). This is an important step to ensure that your object is watertight or manifold. Objects that aren't merged but are overlapping or intersecting one another will cause problems when you try to print the model later.

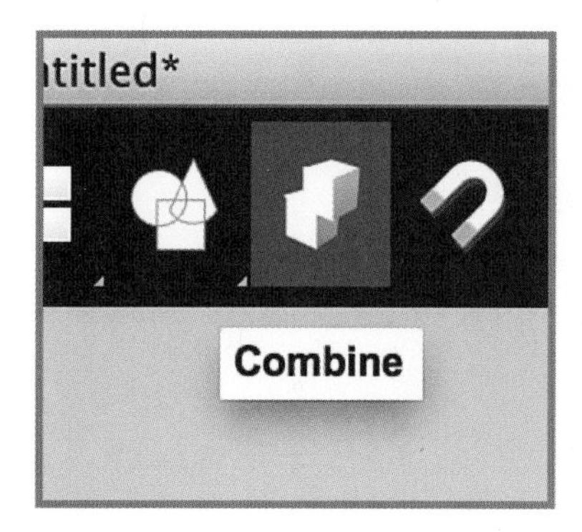

FIGURE 7.17 Combine tool.

The **Snap** tool allows you to quickly move objects and snap them to each other by clicking on one face that you want to snap to another (see Figure 7.18).

FIGURE 7.18 Snap tool.

The **Measure** tool shown in Figure 7.19 pops up a window and enables you to measure multiple objects using a number of object types (faces, edges, vertices, or bodies) and measurement types (distances, angles, areas, and volumes). The measure window is movable by dragging it, or you can collapse it to the side. You need to scroll the little window down to see the results of your measurement choices.

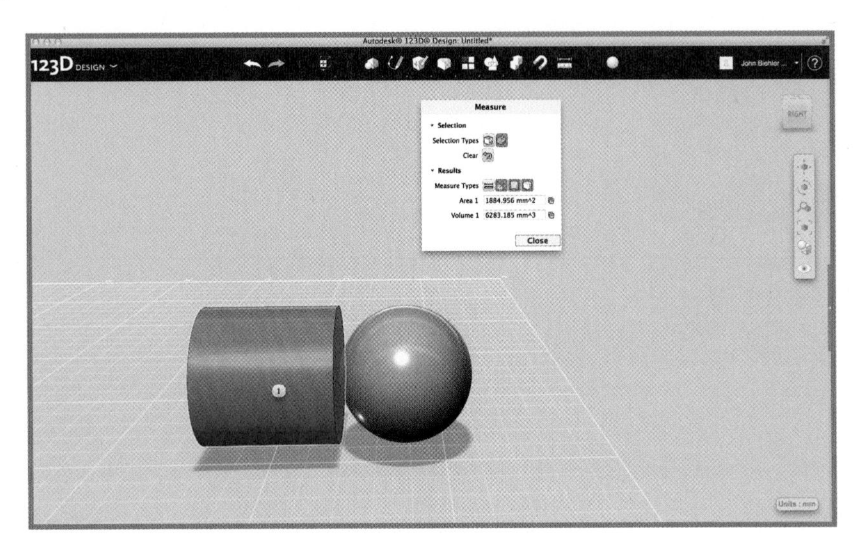

FIGURE 7.19 Measure tool.

The final menu option is for **Materials** (see Figure 7.20). This tool enables you to change the visual appearance of a specific object. These materials include various types of metals, woods, plastics, and similar raw materials in various shades and colors. It doesn't change anything other than the physical look of the object, but it can be useful if you want to visualize the object as it may appear after you've printed it in 3D. You can also change the colors of objects, which can be helpful in keeping things organized while you're creating.

TIP

These colors won't carry over to your 3D printed objects because the materials used for 3D printing will dictate the object's color. However, these colors can help you visualize the finished product.

FIGURE 7.20 Materials options.

The final options on the top menu are for your account management and help. This menu enables you to view your account, change your password, and view your current projects on the Autodesk website. These options will take you out of the 123D Design program and launch a web browser.

The help section is a summary of quick tips, provides some video tutorials on how to use certain tools and options, and contains links to the community forums for 123D Design.

On the right side, directly below the **Account/Help** menu options, is the **View Cube**. This icon shows you the current orientation of the workplane and has a Home button you can press to restore the view to the default one. There is also a drop-down button on the lower right of the **View Cube** that enables you to choose between perspective or orthographic views of the workplane.

Below the View Cube is the **Navigation Bar**. This set of controls lets you change the camera views using Pan, Orbit, Zoom, and Fit controls. Below these controls are the options to display your object in various forms, including outline mode. You can also turn off the materials you might have chosen earlier to just see an outline of the object. This is useful if you are working on a number of objects and want to see how they intersect on the inside. You can also switch between viewing solids and sketches—both or neither as well.

On the far right bottom is a **Units** indicator. Clicking this will enable you to switch the units of measure being used for your objects.

Depending on the object selected, an Attribute submenu shown in Figure 7.21, might appear at the bottom of the workplane. This menu contains a number of shortcuts to control your object without having to go up to the main top menu and navigate through the submenus. These shortcuts include **Materials**, **Move**, **Scale**, **Hide**, and **Grouping** options.

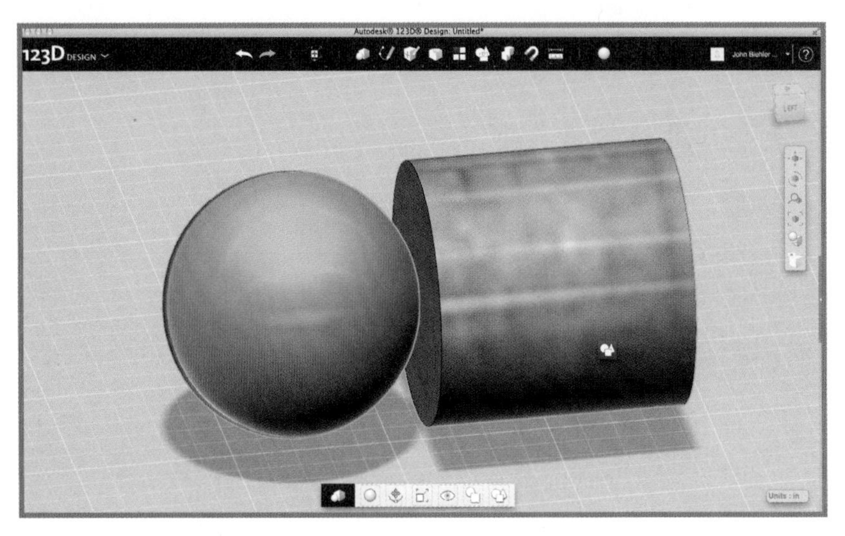

FIGURE 7.21 Attributes submenu.

Summary

This wraps up the tour of the controls, menus, and options for 123D Design. In the next chapter, we start using these tools to create objects in some sample projects to get you started creating your own 3D models.

123D Design Exercises for Mac and PC

Similar to Chapter 6, "123D Design Exercises for iPad," this chapter focuses on a few simple exercises in 123D Design (desktop), but for the Mac and PC. The exercises use some of the tools that were explained in Chapter 7, "Workspace Basics of 123D Design for Mac and PC."

In this chapter, we walk through the steps for creating two basic objects using the tools we covered in Chapter 7. In the first exercise, we create a basic coffee mug; then we'll create a business card holder.

After launching 123D Design, sign in to your AutoDesk account and create a new project (see Figure 8.1).

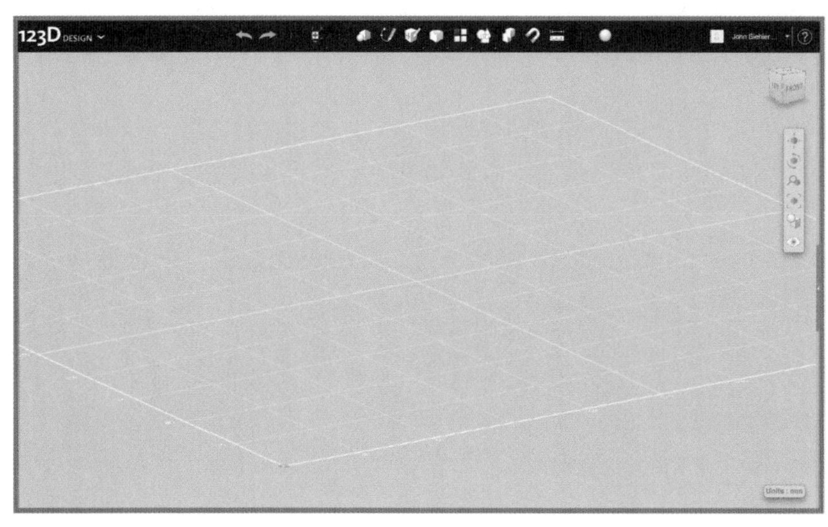

FIGURE 8.1 Blank workspace.

Exercise: Create a Coffee Mug

The first exercise will be to create a coffee mug. The specifics outlined here are to make one type of mug, but feel free to experiment with the sizes and shapes depicted if you'd like. In most cases the details have a lot of flexibility. You can always hit the Undo button if you don't like what you come up with:

1. Starting with a new project and an empty workspace, go to the **Primitives** menu and click the cylinder. It will appear below your cursor. Move it to the middle of the workspace, but don't click the mouse button quite yet to drop it in place (see Figure 8.2).

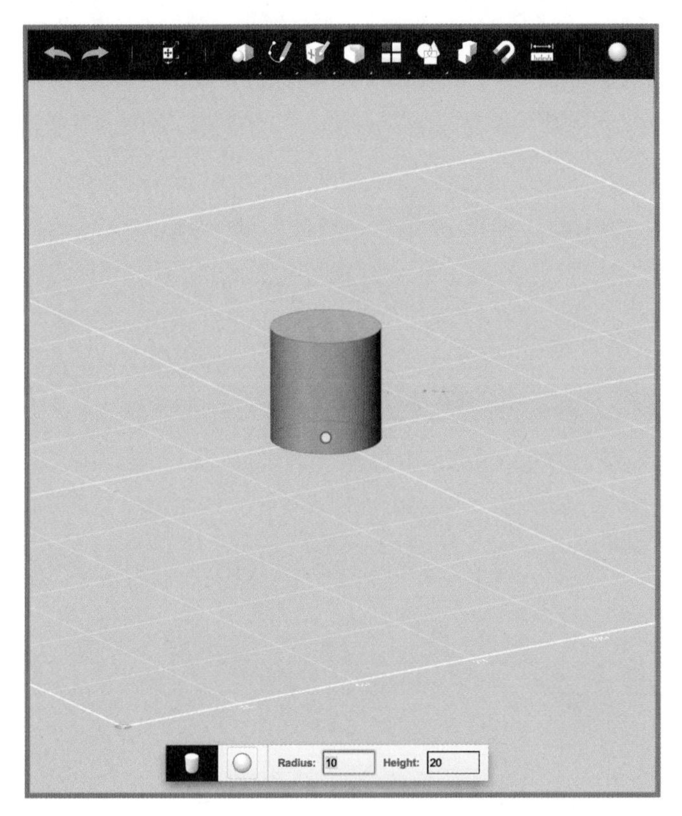

FIGURE 8.2 Setting the cylinder's radius.

2. On the bottom of the screen, you are able to set the radius and height of any primitive you bring down (before clicking the mouse again to drop it in place). For this exercise, set the radius to 15 and the height to 40—the cursor should already be in place so you can just type the numbers (see Figure 8.3).

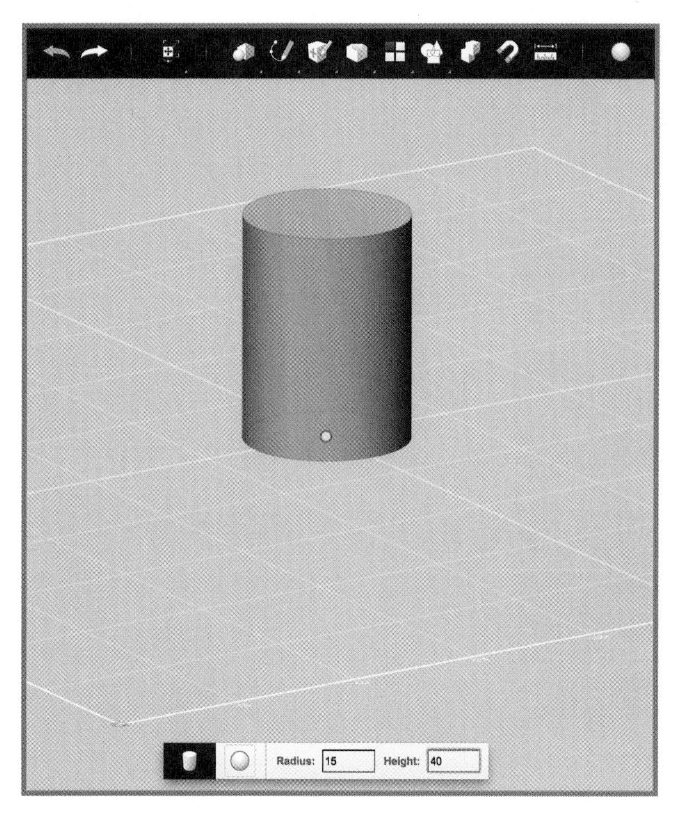

FIGURE 8.3 Resized cylinder.

3. Next, you'll need to zoom in or select the **Fit** option from the right menu to move the camera in close to the cylinder.

TIP

If you're using a mouse, you can use its scroll wheel to zoom in and out—the camera will center on wherever the cursor is located

4. Next, from the **Modify** menu, select **Shell** (see Figure 8.4). We're going to hollow out the cylinder similar to what we did in Chapter 6.

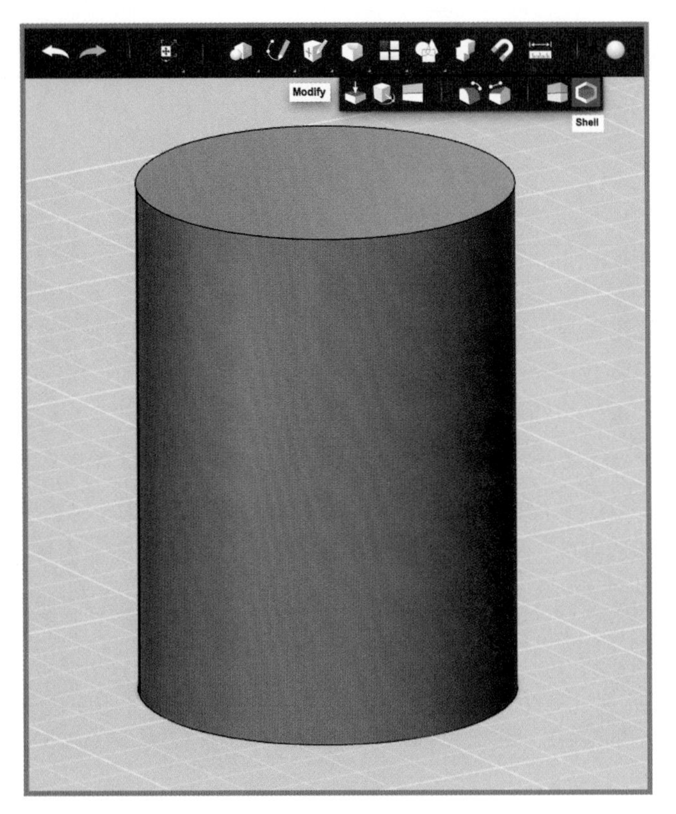

FIGURE 8.4 Selecting the Shell tool.

5. You can uncheck **Tangent Chain** in the window that appears on the bottom (see Figure 8.5).

FIGURE 8.5 Shell tool options.

6. We'll use a thickness of 1 in the exercise, but you can make this coffee mug as thick as you'd like. Try different numbers to see the effect each has on the cylinder's walls (see Figure 8.6). After entering the thickness, click the top of the cylinder; notice the hole that appears (see Figure 8.7).

FIGURE 8.6 Try different wall thicknesses.

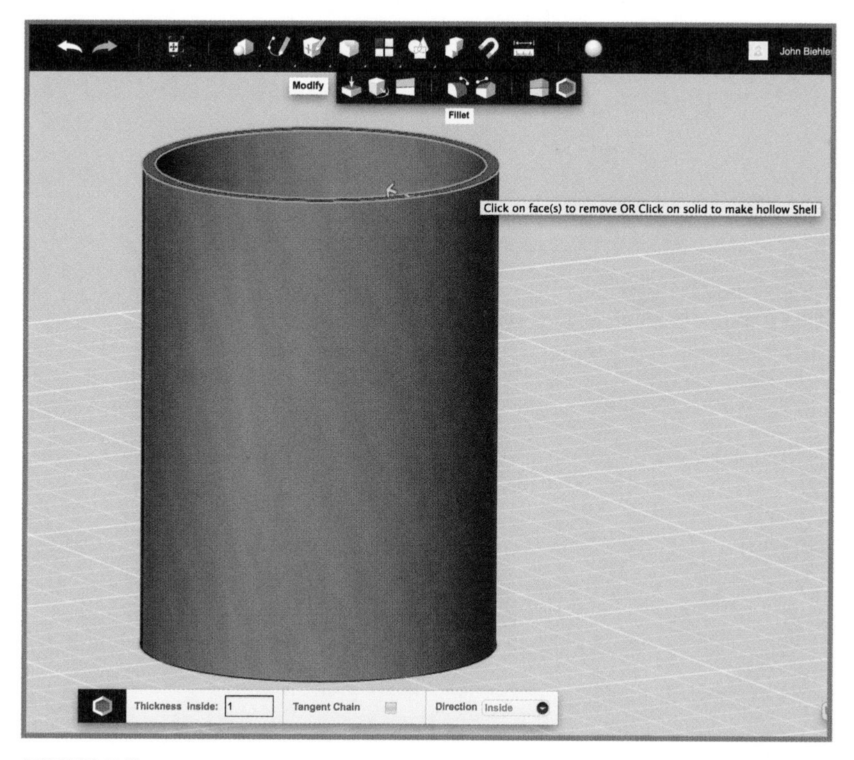

FIGURE 8.7 Shelled cylinder.

7. Now we need to smooth out the top and bottom edges using the **Fillet** option from the **Modify** menu (see Figure 8.8).

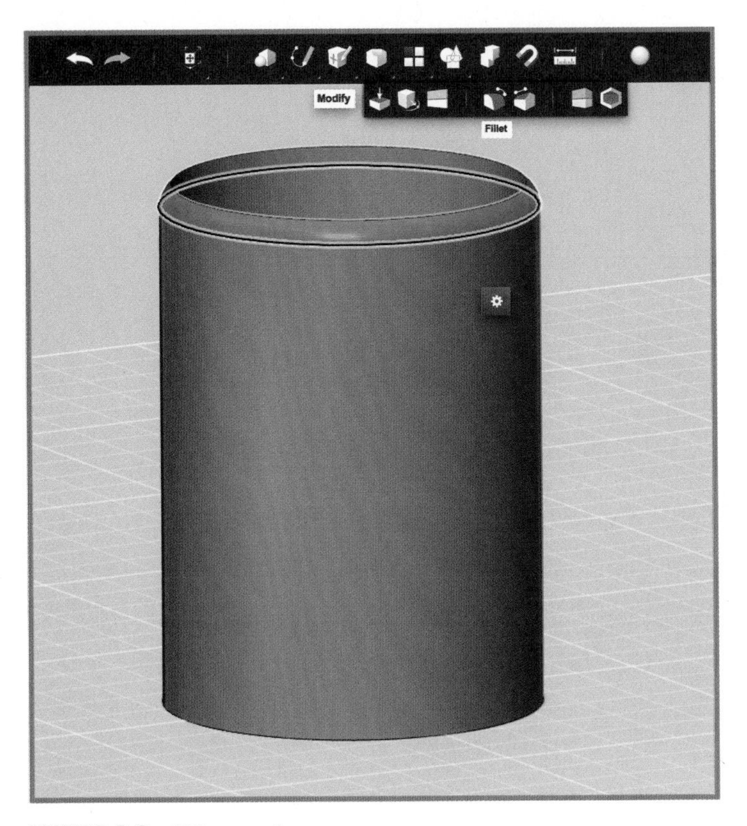

FIGURE 8.8 Fillet tool.

8. Click the outer edge of the top of the cylinder and drag the white grip control to affect the amount of filleting on that edge (see Figure 8.9).

FIGURE 8.9 Filleting the edges.

9. Do the same for the bottom edge.

TIP

Use care with the bottom because if too much filleting is applied, it can be hard to 3D print without support material. You just want to round the edge, not create an overhang as the cylinder goes up from the bottom.

Making the Mug Handle

There is no simple way to intersect a handle with the cylinder in this version of 123D Design, so we'll walk through one way of creating and attaching the handle; there are a number of ways this could be accomplished. Here's one of them:

1. Start by dropping another cylinder onto the workspace (see Figure 8.10) and sizing it 15×40, similar to the original one. Don't worry, this will all make sense in a moment.

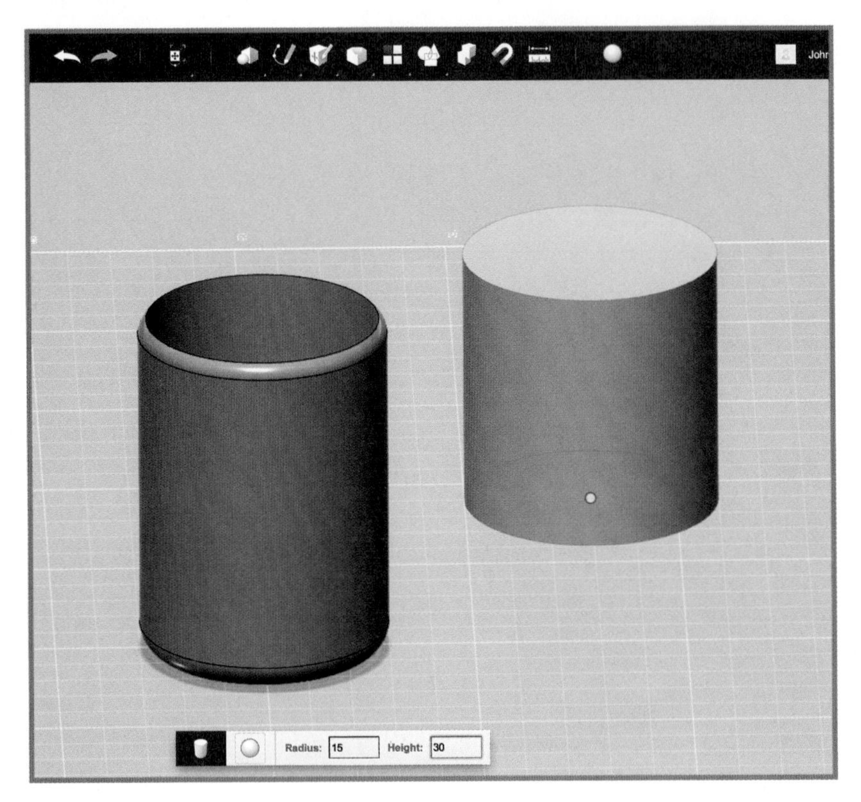

FIGURE 8.10 Adding another cylinder.

2. We'll make the handle out of a torus, so drop one down near the second cylinder you just added (see Figure 8.11).

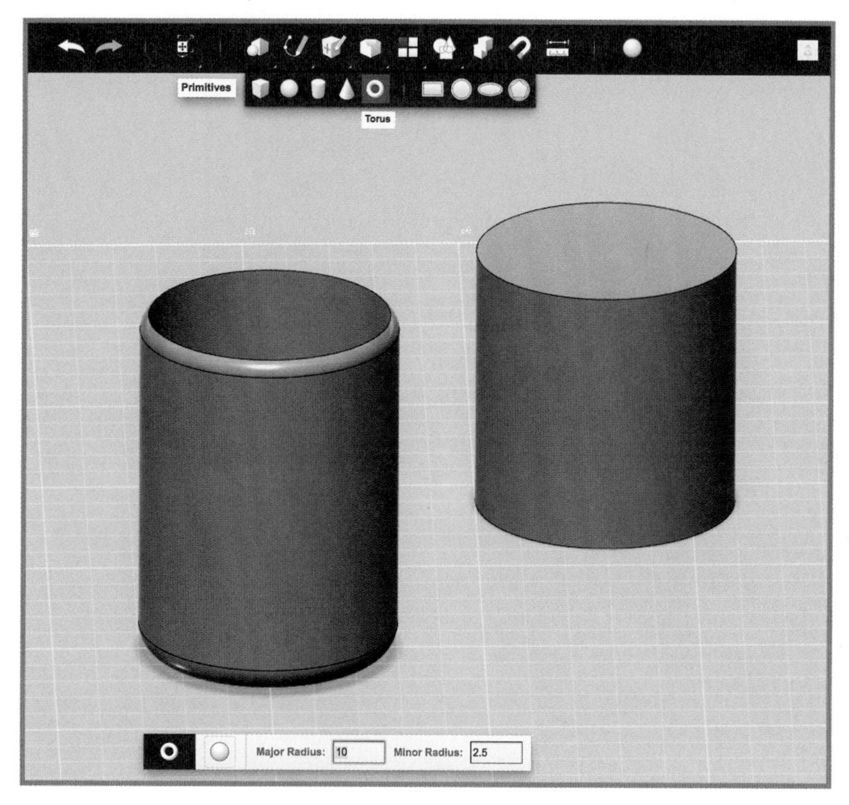

FIGURE 8.11 Selecting the torus primitive.

3. You can size the torus as big or small as you'd like; we'll just use the default size for this exercise, though (see Figure 8.12).

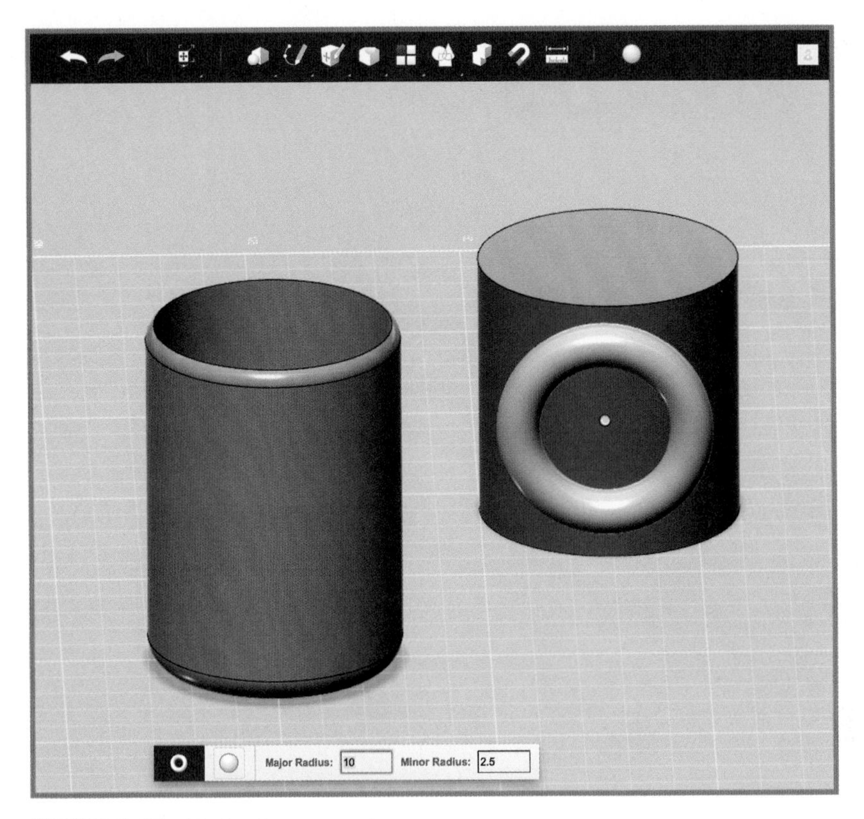

FIGURE 8.12 Default torus size.

4. Move the view cube to the Top view using either the mouse or the view controls on the top right.
5. Select the torus and choose the **Move/Rotate** tool. You're going to rotate the torus 90° so that it intersects the cylinder (see Figure 8.13).

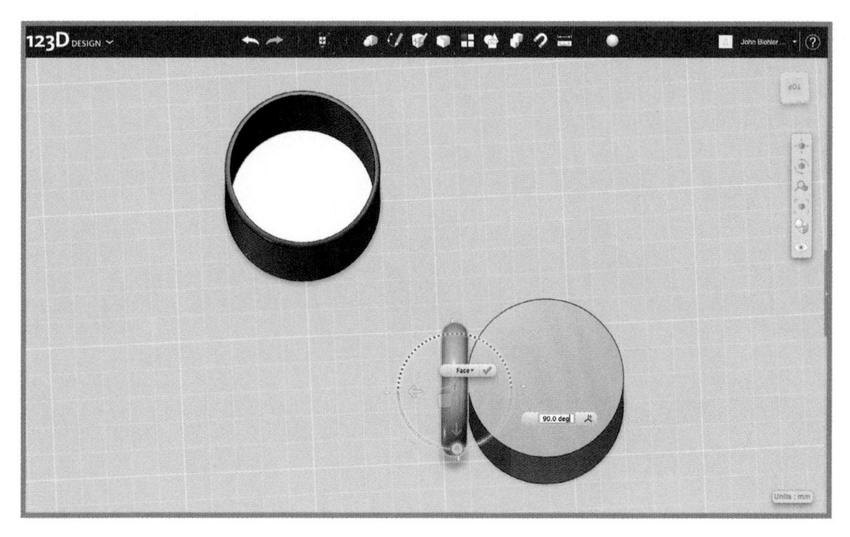

FIGURE 8.13 Rotating the torus.

6. Then move it in or out of the cylinder as desired to form a handle on the cylinder (see Figure 8.14).

FIGURE 8.14 Positioning the torus to make a handle.

7. Now for the fun part. We're going to use the **Combine** tool from the top menu to subtract the cylinder from the torus handle. Be sure the tool is in **Subtract** mode (see Figure 8.15). This will leave behind a trimmed torus that we can attach to the original cylinder as a handle.

FIGURE 8.15 Combining the objects.

8. To use the **Combine** tool, click it and then click the object you want to keep, which is the **Target Solid** in this case. Click the torus first (see Figure 8.16). The **Combine** tool will then switch to the **Source Solid** selector. Click the cylinder, and both objects should now be outlined in blue (see Figure 8.17).

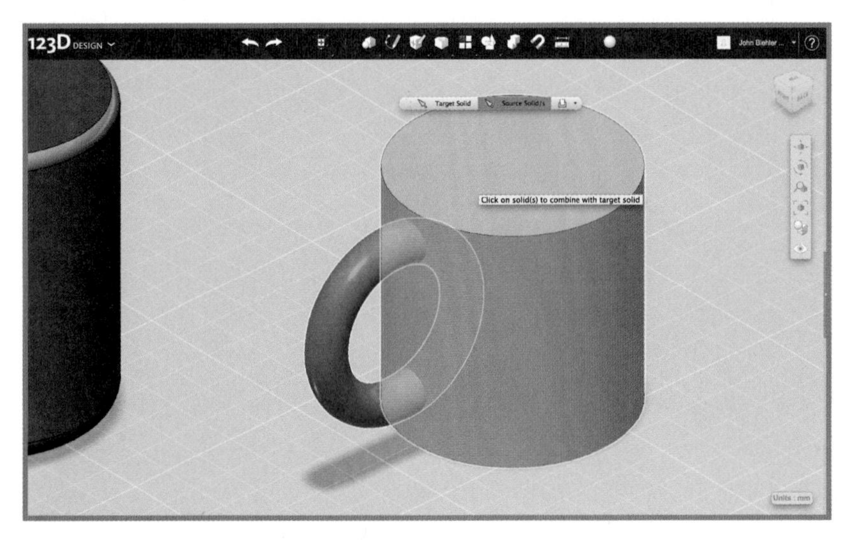

FIGURE 8.16 Click the torus first.

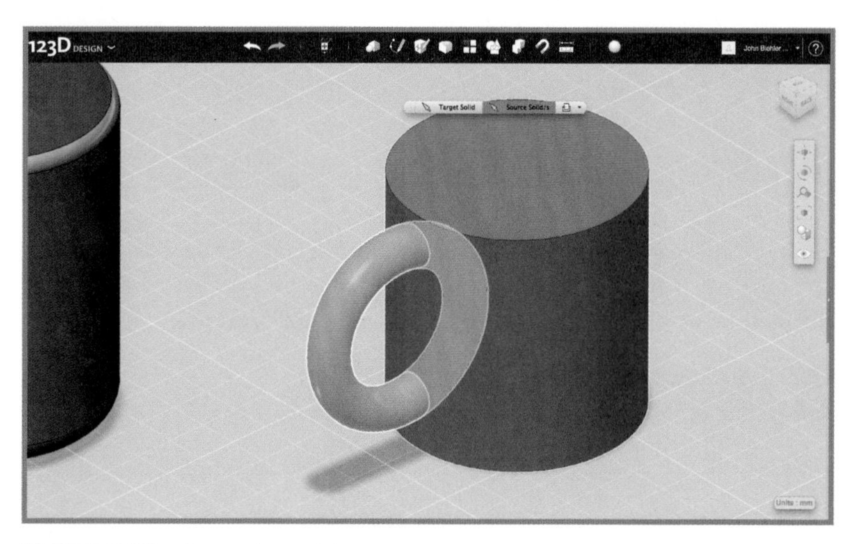

FIGURE 8.17 Selecting objects to combine.

9. Press **Enter** or click the workspace to remove the cylinder and leave behind our handle (see Figure 8.18).

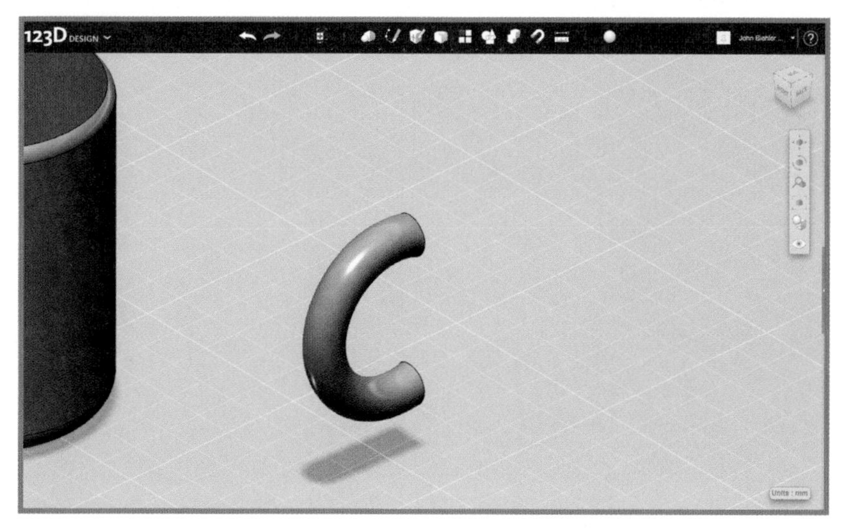

FIGURE 8.18 The mug's handle has been created.

TIP

If you've selected the objects in the wrong order, simply use the Undo button on the top menu to move back a step and select the objects in the desired order.

10. Now move the newly created handle into position on the original cylinder (see Figure 8.19). There are a few ways to create this handle, but this method is the easiest to do using the available controls in 123D Design in the least amount of steps.

FIGURE 8.19 Positioning the handle.

11. You can adjust its height as desired using the **Move** tool's control vertically. Check your model from multiple angles to ensure that the handle is 90° to the cylinder; otherwise, the next step might not work.

TIP

You could also scale the handle at this time if you'd like a bigger or smaller one.

12. The final step is to use the **Combine** tool again to join the two objects together. Click the **Combine** tool (see Figure 8.20) and ensure that the dialog box that appears shows the **Join** option selected (it should be the default). Then click the target and source objects (see Figure 8.21).

FIGURE 8.20 Joining the handle to the mug.

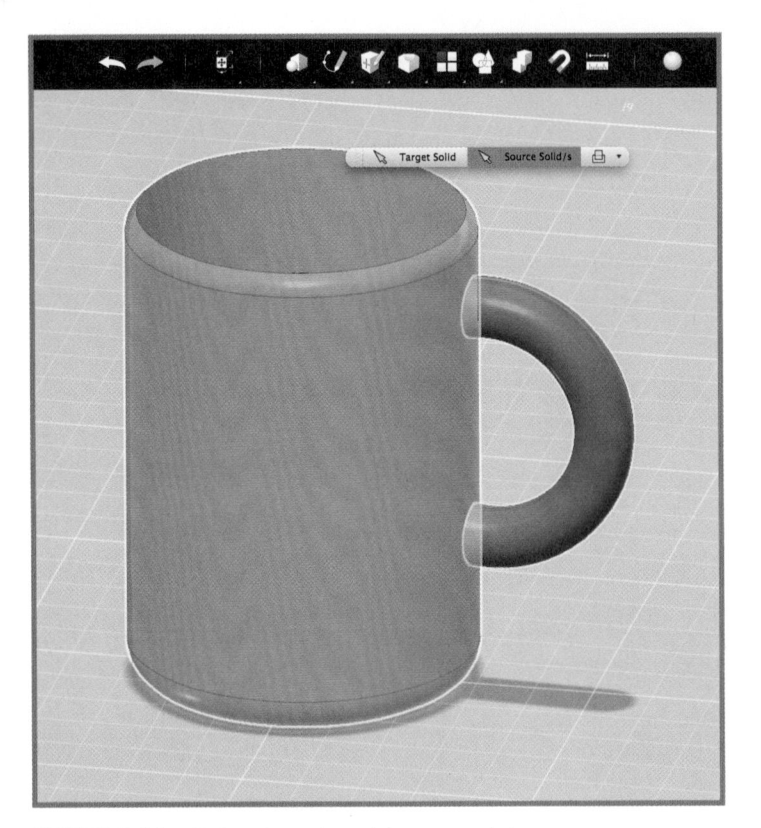

FIGURE 8.21 Selecting the objects to join.

TIP

The order in which you select the objects in step 12 doesn't matter because you're joining both together and not subtracting like you did earlier.

13. When both are blue, click the workspace or press **Enter** to apply the change. This makes a nice manifold model that's ready for 3D printing (see Figure 8.22).

FIGURE 8.22 Completed model.

14. Save your work, and now you can export your STL for printing. The finished mug turned out well (see Figure 8.23).

FIGURE 8.23 Finished printed mug.

Exercise: Create a Business Card Holder

Next up, we'll create a business card holder using some of the other tools in 123D Design:

1. Start a new project (see Figure 8.24).

FIGURE 8.24 Blank workspace.

2. Go to the **Primitives** menu and grab the box.
3. When it appears on the workspace, go to the bottom info box and type in 100 length, 60 width, and 40 height. Press **Enter** (see Figure 8.25).

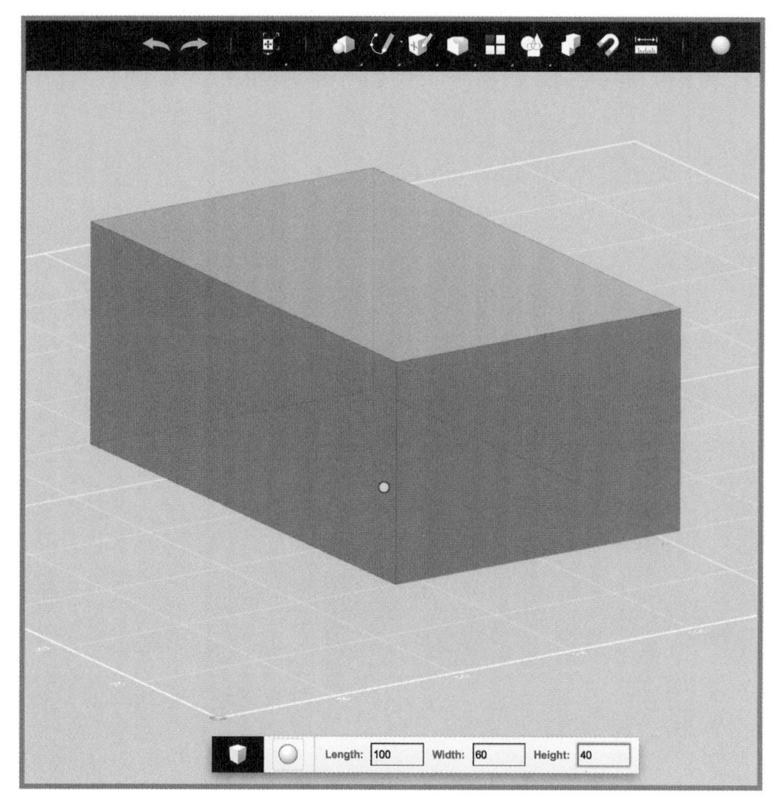

FIGURE 8.25 Starting with a box.

4. Just like the mug earlier, we'll use the **Shell** tool from the **Modify** menu to hollow out the box (see Figure 8.26). Set the thickness to 3 in the bottom information box.

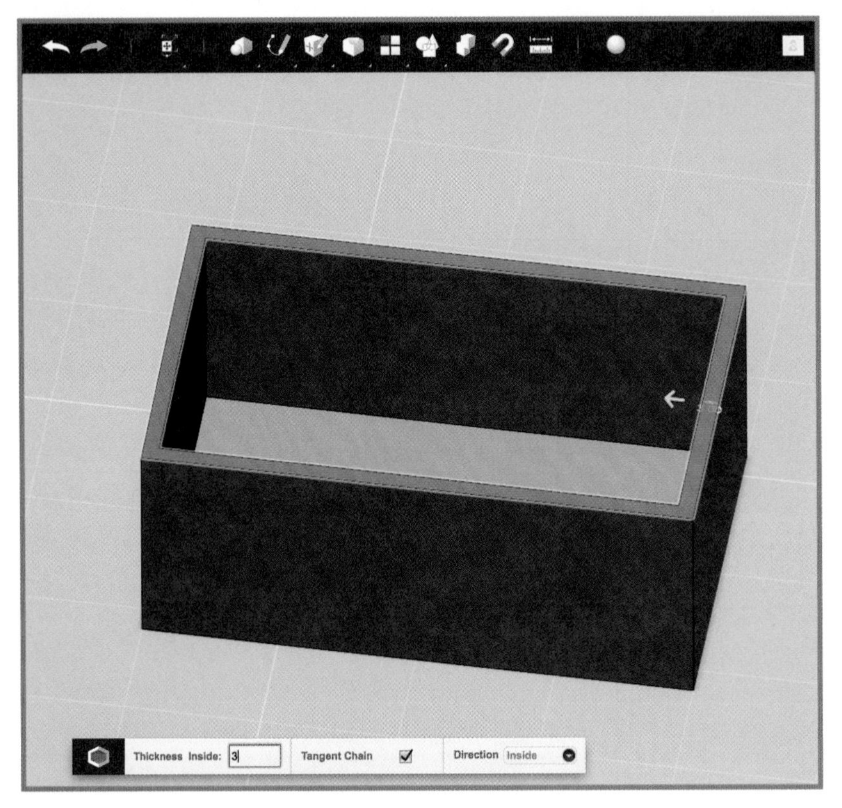

FIGURE 8.26 Hollowing out the box.

5. Next, select the **Chamfer** tool from the **Modify** menu (see Figure 8.27). Then select the front top edge (see Figure 8.28).

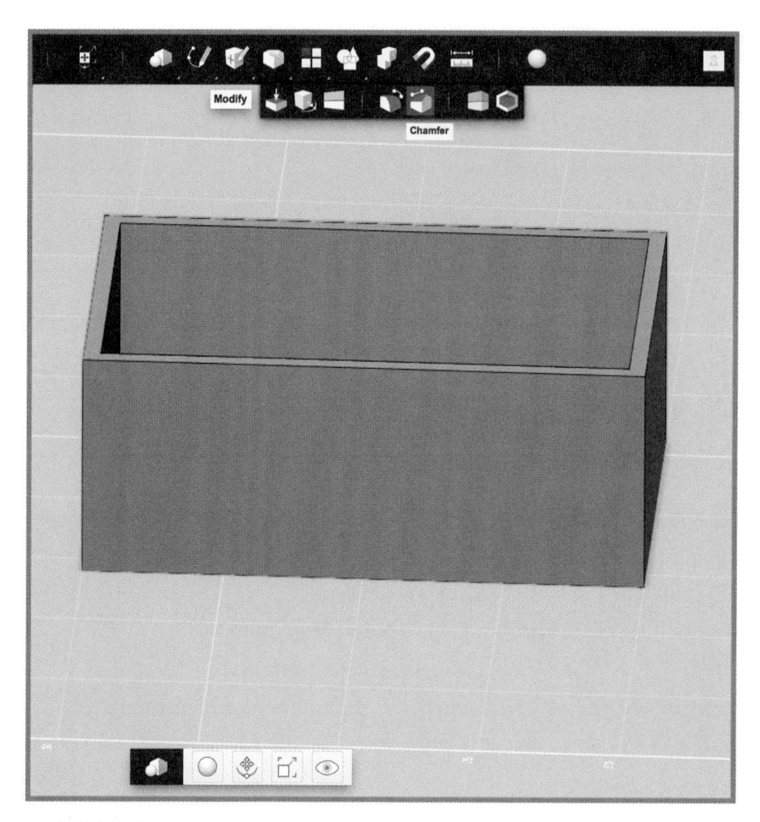

FIGURE 8.27 Chamfer tool.

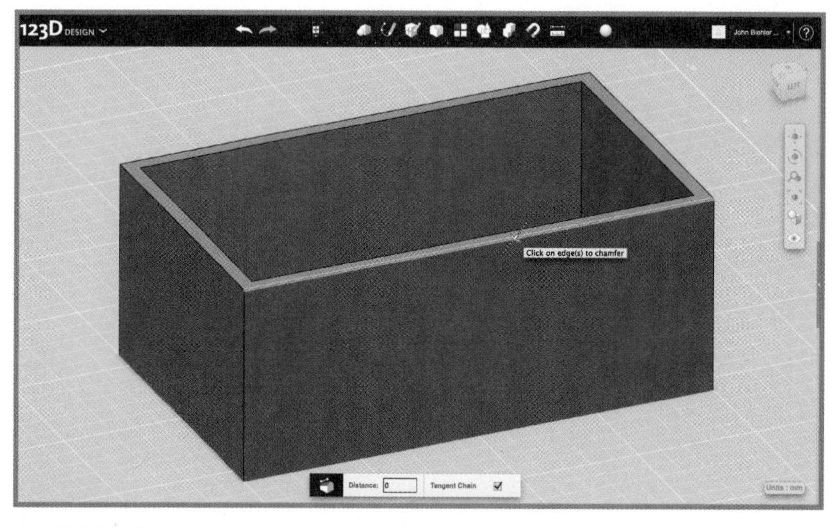

FIGURE 8.28 Chamfering the top edge.

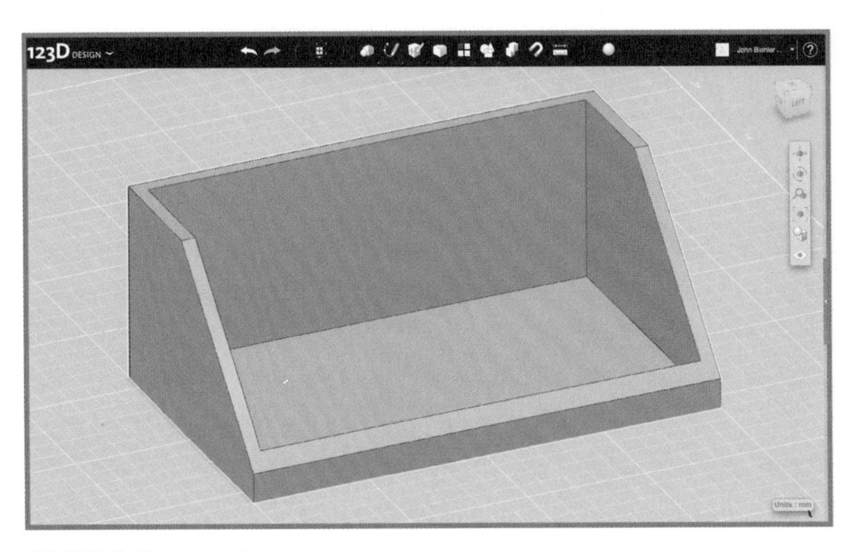

FIGURE 8.29 The box is taking shape.

6. Drag out the **Chamfer** control to get something that resembles Figure 8.29.
7. We're going to use the **Tweak** tool, which is also from the **Modify** menu (see Figure 8.30), to pull out the bottom of the back wall using the default **Extend** submenu option, as shown in Figure 8.31. This provides a nice slope for the business cards to rest against in the holder.

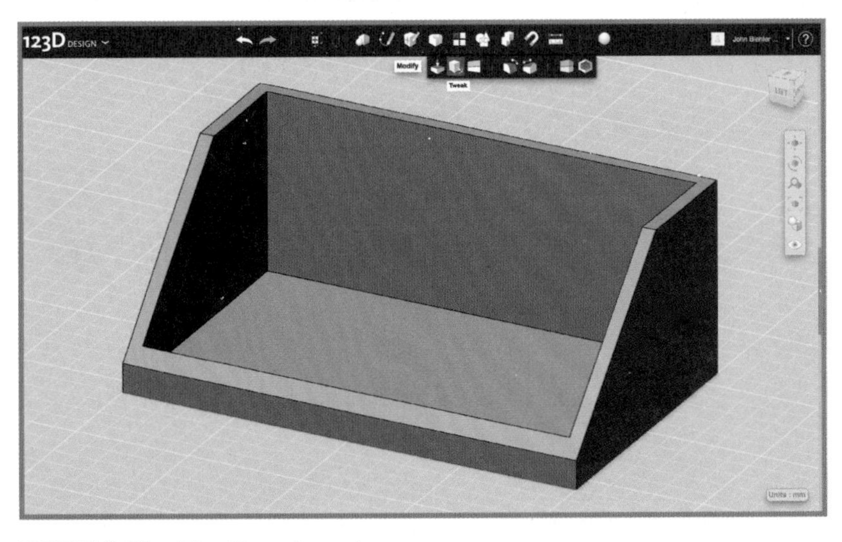

FIGURE 8.30 The Tweak tool.

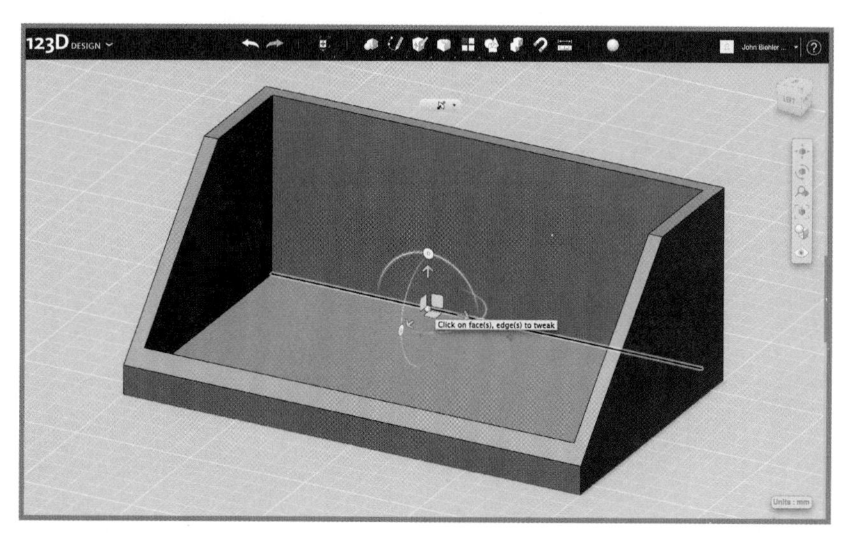

FIGURE 8.31 Tweaking the back wall of the box.

8. Click the bottom edge of the back wall, and then click the white grip arrow facing the front of the holder. You can either drag it out by hand or enter 15mm in the precision box that appears on the right (see Figure 8.32).

CAUTION

If you drag it out manually, ensure your movement is slow and straight because the Tweak tool can move in any direction and will deform your holder if you move left or right while pulling out.

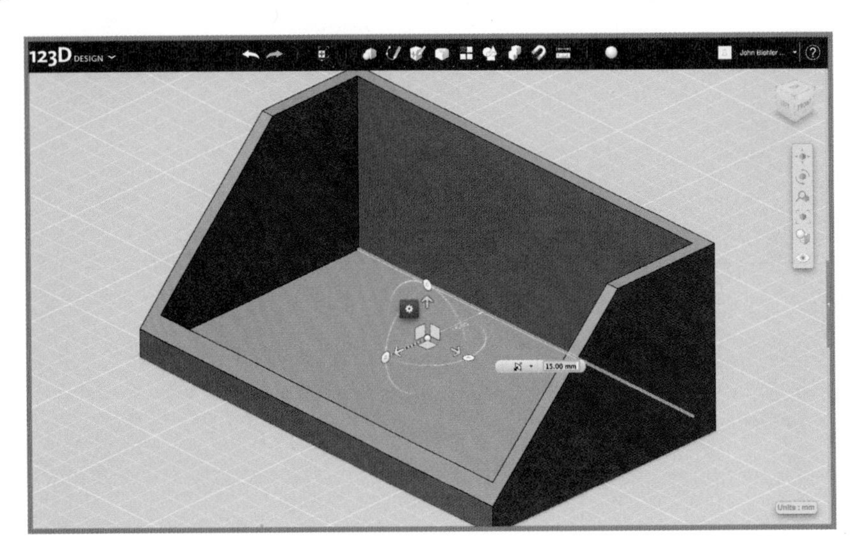

FIGURE 8.32 Precision sizing the tweak.

9. Now you should have a nice rear slope in the holder (see Figure 8.33).

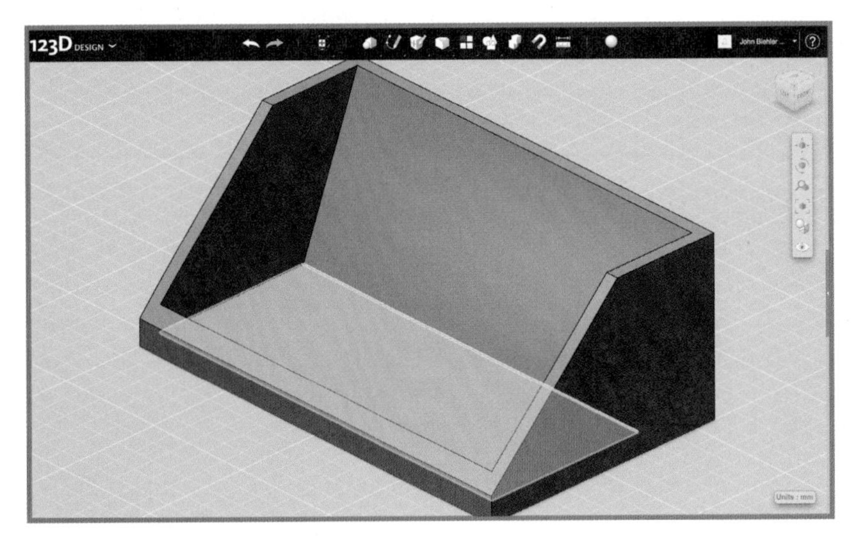

FIGURE 8.33 Rear wall is now sloped.

10. To finish it, head back to the **Modify** menu and select **Fillet** (see Figure 8.34).

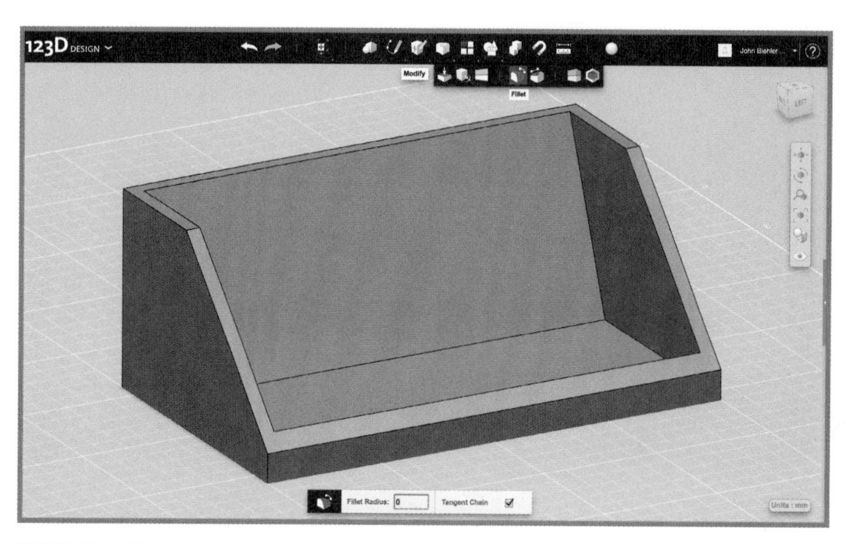

FIGURE 8.34 Finishing with a fillet.

11. With the **Fillet** tool selected, click the right edge of the holder (see Figure 8.35) and drag out the Fillet as desired to round out the edges, as shown in Figure 8.36.

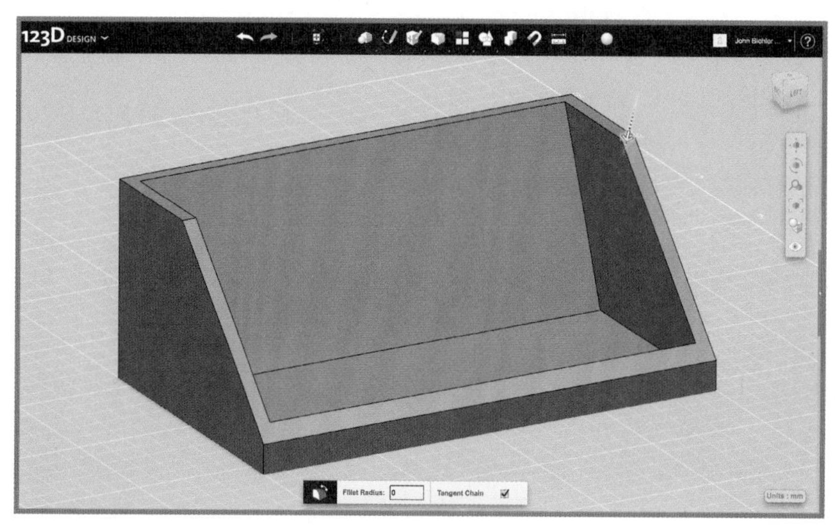

FIGURE 8.35 Start with right edge.

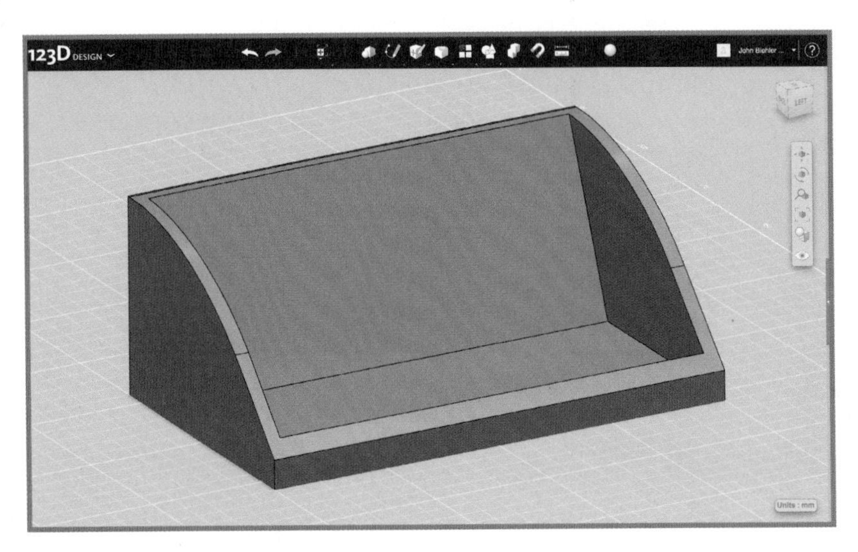

FIGURE 8.36 Nice rounded edges.

TIP

As you drag, it will eventually lock and duplicate the fillet on the left side as well, giving you smooth and uniform filleted sides for the holder.

12. All done! Save your work and export your STL for printing (see Figure 8.37).

FIGURE 8.37 Printed business card holder.

Summary

In this chapter, we created a coffee mug and a business card holder using the tools in 123D Design. The tools are a mix of roughing out the shapes and finishing tools that can really make your designs look polished.

Now that you have made a number of models in these exercises, in the next chapter we'll cover some tips to get them ready them for 3D printing.

Preparing 3D Models for Printing

Now that you've had some experience creating 3D models, what do you need to be aware of before printing them on your 3D printer? This chapter covers a number of things to ensure your model has the best chance for being printed on a typical desktop 3D printer successfully. Because all 3D printers are a little different from one another, bear in mind these tips might not apply to your specific machine, but they are provided as general information for the current, most common models on the market at the time of this book.

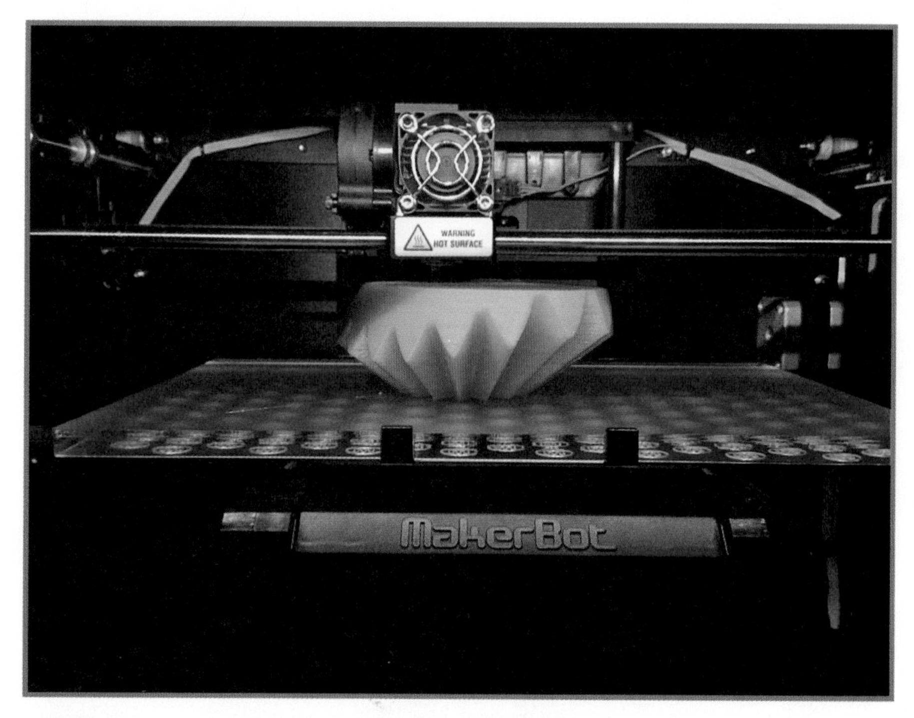

FIGURE 9.1 Printing with a MakerBot Replicator 2.

We'll break this chapter into three sections: scale and dimension, manifold geometry, and orientation. First, let's describe what's happening when you want to print a 3D model.

Code Used for Printing a 3D Model

All current desktop 3D printers use standard g-code to operate. This code is a series of text-based instructions that tell the printer where to move the print head and/or platform (depending on the type of printer). In conjunction with this, the code tells the printer how fast to deposit the material and the temperatures that are required. These sets of instructions are typically generated by software commonly referred to as a *slicing program* or *slicer*.

> **NOTE**
>
> These slicing programs vary by printer manufacturer and are usually included with the printer or available to download from the Internet.

Some are printer-specific, such as Cura (for Ultimaker) or MakerWare (for MakerBot), while others can be configured for a variety of machines using different profiles such as Replicator G, Repetier Host (see Figure 9.2), and Slic3r.

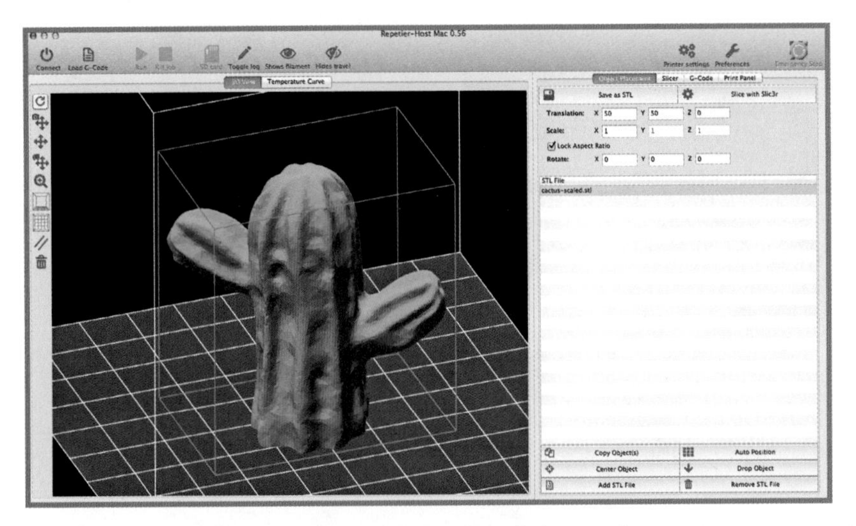

FIGURE 9.2 Repetier Host printing and slicing software.

These slicers generally create text files with the various instructions for use by the 3D printer. The text files can either be used on the computer directly connected to the 3D printer or be saved onto memory cards or USB thumb drives that are then inserted in printers that have the ability to print directly from these storage devices.

The files are usually fairly small but can be as much as a few hundred megabytes of data depending on the complexity and scale of the model being printed. An example of the contents of a g-code file is shown in Listing 9.1.

LISTING 9.1—Sample G-Code Output with Configuration Settings for the Printer

```
; layer_height = 0.3
; perimeters = 3
; top_solid_layers = 3
; bottom_solid_layers = 3
; fill_density = 0.1
; perimeter_speed = 10
; infill_speed = 15
; travel_speed = 40
; nozzle_diameter = 0.4
; filament_diameter = 1.75
; extrusion_multiplier = 0.898
; perimeters extrusion width = 0.44mm
; infill extrusion width = 0.44mm
; solid infill extrusion width = 0.44mm
; top infill extrusion width = 0.44mm
; first layer extrusion width = 0.70mm

G21 ; set units to millimeters
M104 S185 ; set temperature
M92 X185.35 ; calibrate X
M92 Y194.281 ; calibrate Y
M92 Z2267.720 ; calibrate Z
;G1 X0 F3000 ; home X axis
G1 Y0 F3000 ; home Y axis
;G1 Z5 F5000 ; lift nozzle
M109 S185 ; wait for temperature to be reached
G90 ; use absolute coordinates
G92 E0
M82 ; use absolute distances for extrusion
G1 Z0.350 F2400.000
G1 F1800.000 E-3.00000
G1 Z0.850 F2400.000
G92 E0
G1 X45.049 Y45.148
G1 Z0.350
G1 F1800.000 E3.00000
G1 X45.539 Y44.708 F450.000 E3.05470
G1 X48.019 Y42.898 E3.30970
G1 X48.649 Y42.508 E3.37124
G1 X49.319 Y42.228 E3.43156
```

```
G1 X49.849 Y42.118 E3.47652
G1 X52.879 Y41.728 E3.73025
G1 X53.299 Y41.698 E3.76523
G1 X54.209 Y41.828 E3.84158
G1 X55.049 Y42.198 E3.91781
G1 X55.749 Y42.788 E3.99385
G1 X57.999 Y45.378 E4.27880
G1 X58.309 Y45.768 E4.32018
G1 X58.599 Y46.298 E4.37036
G1 X58.809 Y46.828 E4.41771
G1 X58.919 Y47.398 E4.46593
```

These tools look at your model and slice it up in thin layers—not unlike an MRI scanning machine—to determine on a layer-by-layer basis what it needs to do at a given moment during the printing operation. This code contains movement directions for all the involved axes of the printer and the required temperatures for the bed and print head.

Some slicing programs also give you a visualization of the path that the 3D printer will take to create the model, as shown in Figure 9.3.

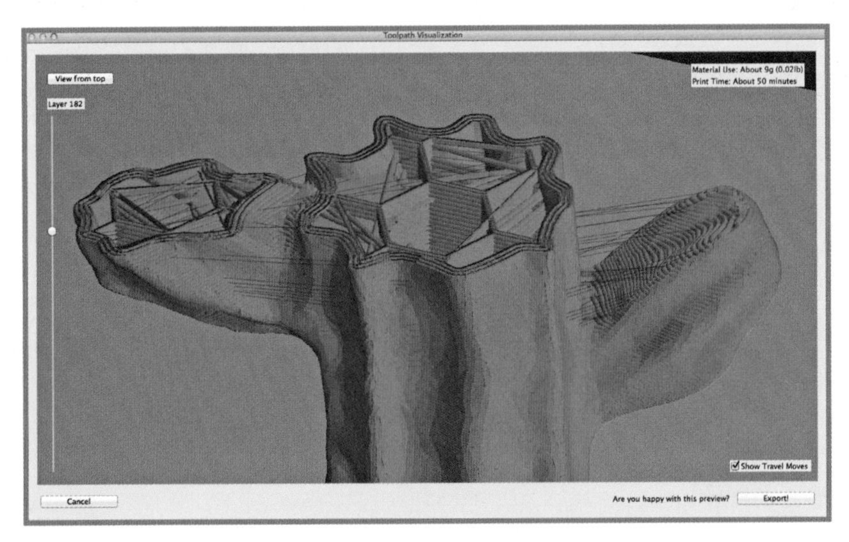

FIGURE 9.3 Toolpath preview of the cactus model from an earlier chapter.

This can be helpful to see whether the software has properly interpreted your model for printing and where issues, if any, may crop up. This is also helpful in understanding how the model will be printed because you can scroll through each layer and see how it will look.

In Figure 9.3, the cactus model in an earlier chapter is previewed at the 182nd layer during printing with MakerBot's MakerWare software slicing tool. It also provides estimates of the amount of printing material used and the time to print the object with the settings used to slice the model.

NOTE

If you are using a third-party printing service, you won't have to slice your model because the service will handle that for you after you upload your model to it. You'll still need to take into consideration these items in your design because these services can sometimes be very strict and will reject your model if it doesn't conform to the requirements of the service.

→ *See Chapter 16, "Using a Third-Party 3D Printing Service Bureau," for more details about using a printing service.*

Scale and Dimension

Depending on where you live and or where you went to school, the dimensions of an object you design will be either in metric (millimeters, centimeters, meters) or in imperial units (inches, feet, yards) of measure. Some 3D modeling software allows you to change your units of measure to either type.

Currently, most, if not all, 3D printers express their units in metric units, typically in millimeters. The same can be said for all the plastic filament used in these printers, which is either 3mm or 1.75mm feed stock.

Everything is metric, from the speed of the printer, which is typically expressed in millimeters per second (mm/second), to print resolution (a.k.a. layer height), which is expressed in microns (200 microns is equal to 0.20mm, for example).

This can cause problems when you load your model into a typical slicing program (MakerBot's MakerWare or open-source packages such as Slic3r or Replicator G) if you used imperial units when you designed your model.

CAUTION

The problem begins when it tries to convert from imperial to metric (inches to millimeters) as the scale of the object can be off. This can lead to parts that have been carefully measured to fit together or integrated with other objects (a circuit board enclosure, for example) not fitting.

Some, but not all, software slicers will prompt you to resize the object (see Figure 9.4), which might or might not produce a desirable result. In some cases, the software will import the model in the wrong unit of measure (mm instead of cm) so it will appear very tiny, or the opposite will happen and it will be very large.

FIGURE 9.4 Scaling a large model.

The easiest way to avoid conversion headaches is to ensure your measurements are done in the same units all the way down the line, from creation of the model to slicing and printing the object.

TIP

To be safe, always use millimeters because most software can handle it correctly and it's the common format when describing a 3D model.

Manifold Geometry

Sometimes what looks like a simple model on the surface can confuse the slicing software and cause problems when printing. To ensure your models have the best chance of printing successfully, you must ensure they are *manifold*. Often, this also is referred to as being *watertight*. You need your model to be completely sealed and consistent all around. Non-manifold models have parts or faces that intersect. This can cause the slicing software to behave strangely when it generates the g-code including:

- Non-manifold models can cause extraneous printing inside the model that can actually weaken the printed object because the model isn't properly merged together as a single, cohesive object.
- Overlapping or internal faces within your model can cause delays during printing because the slicer will be calculating extra paths for printing that aren't necessary, taking much longer to calculate and print with no benefit.

This can be something as simple as two parts or faces that appear to be touching but aren't actually connected, as shown in Figure 9.5. When printed, this will likely be a failure point in the object as the two faces would be treated as separate objects and not actually merged together.

The slicing software might try to merge parts of the objects together in an effort to streamline the toolpath it generates. Often, however, only the exposed faces will actually appear merged, while inside, the model is a mess and will likely fail easily.

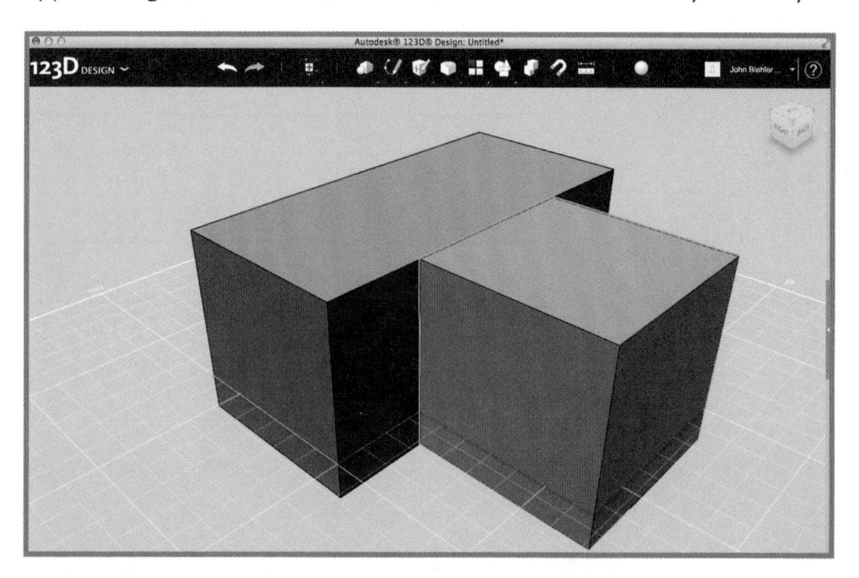

FIGURE 9.5 Simple model with faces that appear to touch.

Many software applications that can create 3D models have tools to verify that the model is manifold or to at least identify problem areas. In some cases, applications will try to repair the model. Sometimes this process can be embedded in the save routine when you export your model.

CAUTION

The slicing software tries to interpret the model as best as it can, but it doesn't always work. Depending on the problem, it will sometimes report errors during the slicing process. These errors can be more warnings than actual errors, but it's often hard to tell until you actually try to print the model.

Orientation

When you design a model, you might need to orient it in such a way that can make it difficult to know which way is up or which is the best way to print it. You can change the model's orientation in either the program used to create it (before saving or exporting the model) or within the slicing program when preparing to print it.

TIP

Not all slicing programs have the ability to rotate the model for printing, so you should be aware of this before sending the model to the slicing program.

Some models can have an obvious "up" orientation, such as the cactus in Figure 9.6, but your model may not.

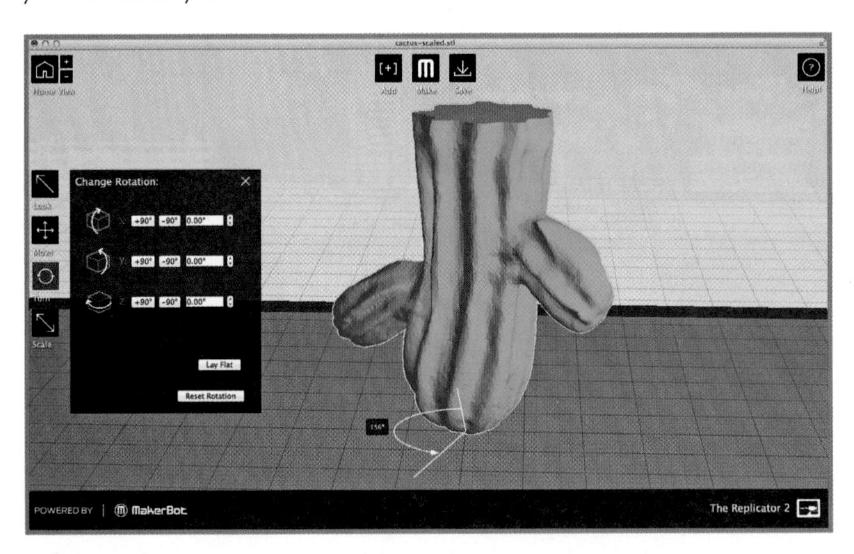

FIGURE 9.6 Changing the model's orientation in MakerWare.

When printing your model using a 3D printer, you will need to understand how a model will be printed for a number of reasons, which are explained in the following sections.

Adding Support Material

One reason is to ensure that the bottom of your model has a flat surface so that, as the printer is depositing material, it has a solid surface to adhere to the bed and won't get knocked loose during the printing process.

If your model is round or has a very small face that is primarily going to be printed against the print bed, you might need to use support material to ensure a solid connection to the print bed during printing. The process of printing can cause your object to literally vibrate off the print bed as the printer moves around unless it is securely attached.

Most slicing programs have an option to add support material (see Figure 9.7). This will add material that is similar to scaffolding and supports the object while it is being printed. This support material will vary depending on your 3D printer, but it is commonly the same material from which your object is printed.

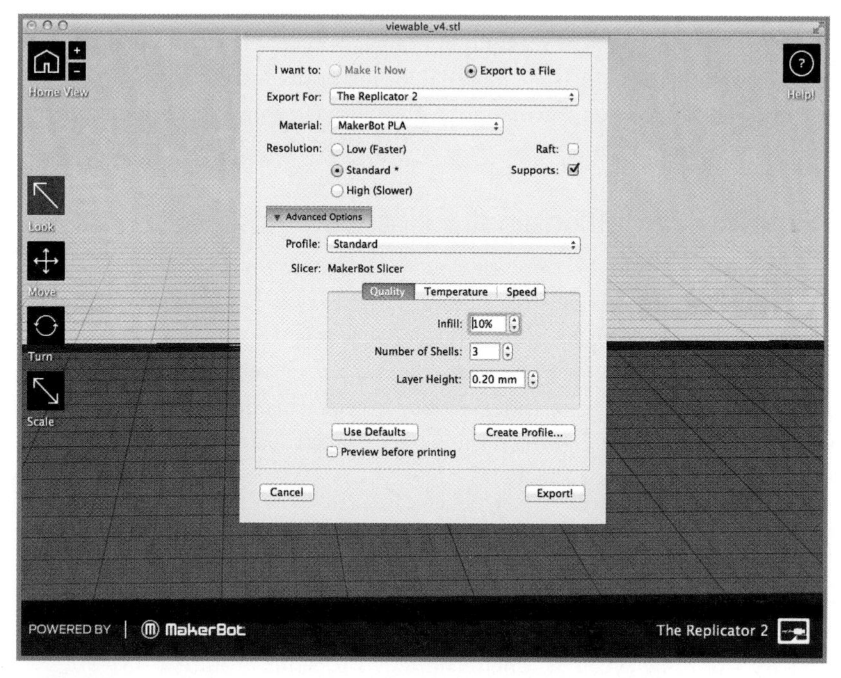

FIGURE 9.7 Enabling support material in MakerWare.

When the printing is done, removing the support material is usually a matter of breaking it off your object, depending on the complexity of your model (see Figure 9.8). Due to the way it's created, it should just come off with little force, like Velcro, from your object. This will vary depending on the type of material with which you are printing.

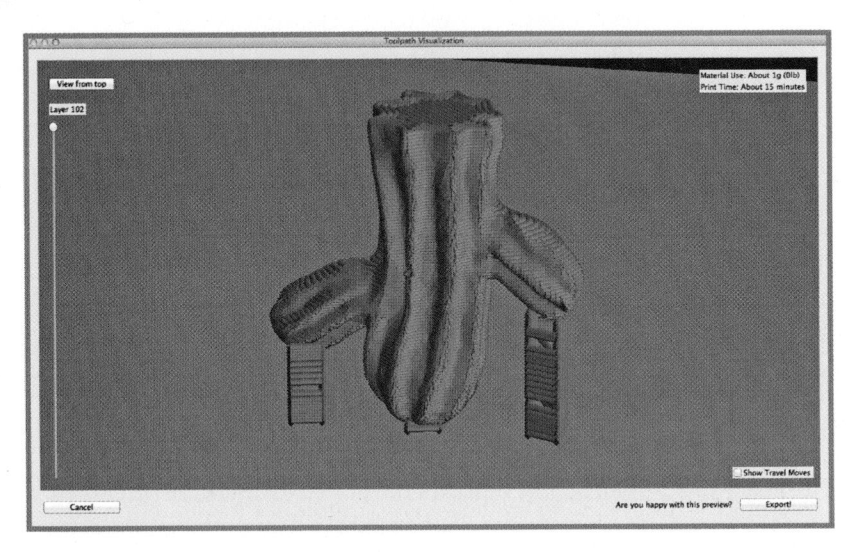

FIGURE 9.8 Printing the cactus upside-down with support material enabled.

Complex models can require more post-processing of the support material, and you might need to use various tools to remove the material. Some printers support the use of secondary extruders and dissolvable support materials that dissolve in water or some other liquid. Consult your printer's manufacturer for the specifics of your machine.

Place Good Side of Model Against Print Bed

You might not want to print the model face up or down, depending on how the object will be attached to the machine's print bed during printing. There may be an obvious good side of the model that you want to be as clean as possible because it is the side that will be seen. If that good side is flush against the print bed, it can be adhered to well during the printing process but appear different from the other sides of your object. This depends on the type of bed your printer has and the means by which the object is attached to the bed (kapton or blue painter's tape, glass, or similar surfaces).

Consider Print Orientation

Another reason to consider print orientation is that some models simply print better one way versus another. This will vary greatly depending on your model (and the printer). Generally, if a model isn't printing properly, one technique is to simply rotate it 45°–90° along the Z-axis and try again (see Figure 9.9). This will change the paths that the printer uses to create the object. If you've had trouble, it can greatly improve your chances of print success.

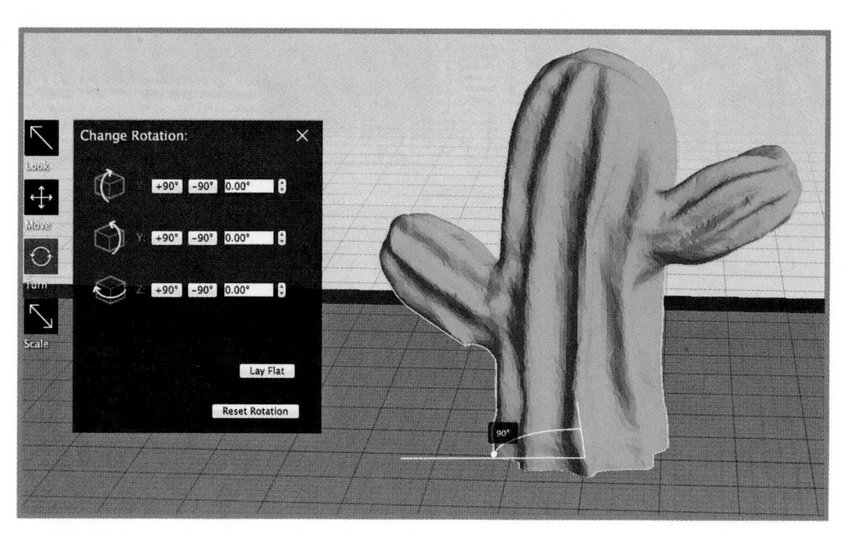

FIGURE 9.9 Rotating on the Z-axis.

Often print failures are caused by the slicer creating paths for the machine to travel that are not optimal depending on the shape of model. By rotating the model, even slightly, the slicing routines have to completely recalculate the toolpath—and that can result in a more optimal print.

When the slicing programs have completed their processes and generated your g-code, your model will be ready for printing.

Summary

In this chapter, we covered some useful techniques to give your model the best chance of successfully printing. With 3D printing, trial and error is sometimes necessary to achieve the best results.

The next section of the book is intended for those with some experience with 3D CAD software. But don't let that scare you off; there are plenty of nuggets of information that will help you as you learn more about 3D printing.

The Difference Between Surface and Solid Models

There are two and a half kinds of 3D models in the CAD world...

NOTE

We won't cover all the gory details of creating 3D CAD models in this book because we assume you already know how, and besides, if we covered each type of 3D model in the various CAD programs, then this book would require a fork truck to move it. If you do need to learn how, a number of resources are available including books, online classes, night school, and so on.

That being said, it is important that you understand the differences between the types of 3D CAD models because choosing the right type of model can mean the difference between producing a 3D print fairly easily versus not being able to produce one at all.

We'll start with a brief description of the major characteristics of 3D solid CAD models and then move on to cover the older 3D surface and mesh types of models.

The Solid Facts About CAD

A CAD solid is a true 3D "solid." If it exists onscreen, then it can exist in the real world. It might be difficult or impossible to manufacture by traditional means, but 3D printing does move the boundaries somewhat.

You can perform mass analysis functions on a 3D CAD solid to determine mass, volume, surface area, center of gravity, moments of inertia, and inertial axes. Figure 10.1 shows a cast aluminum rocker arm from a machine, while Figure 10.2 lists its mass properties.

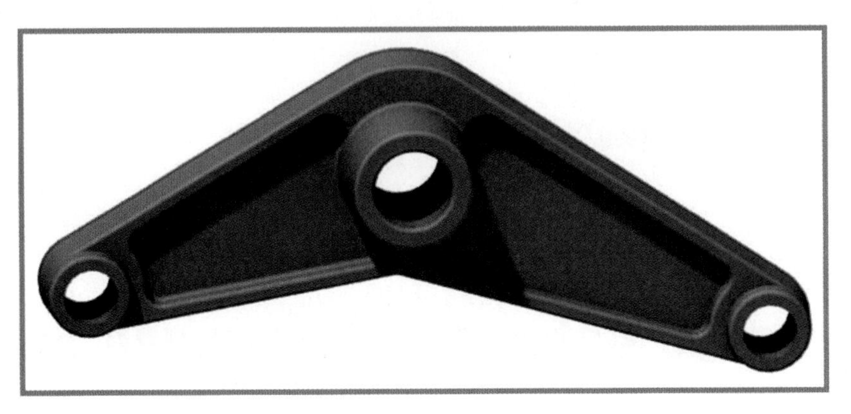

FIGURE 10.1 A typical machine rocker arm.

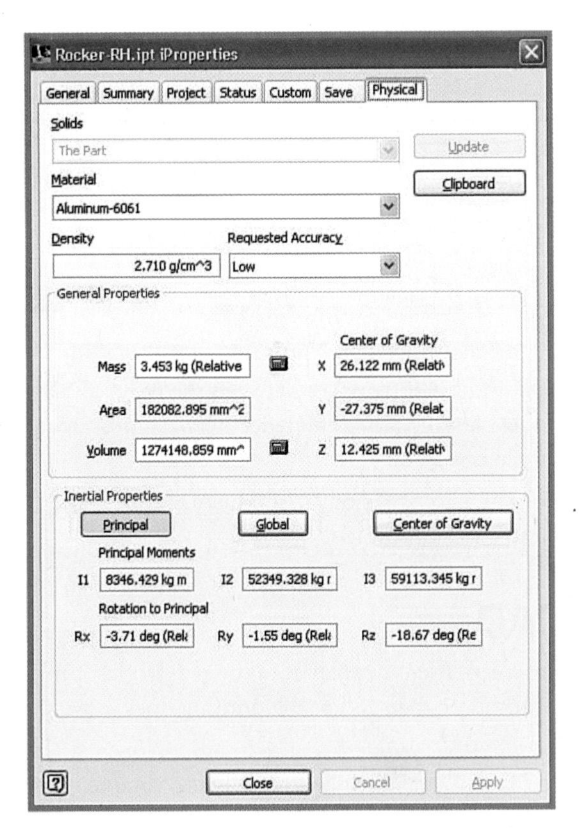

FIGURE 10.2 The mass properties of the rocker arm from Figure 10.1.

The Boolean operations of Union, Subtract, and Intersect can be performed to produce a single complex solid from two or more simpler solids. Figures 10.3, 10.4, and 10.5 show the sequence as a base feature is created, lumps are added, holes are subtracted, and so on.

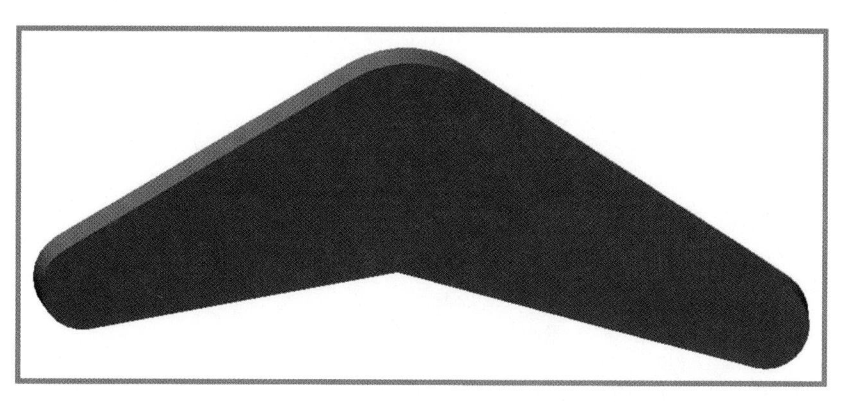

FIGURE 10.3 The base feature for a machine rocker arm.

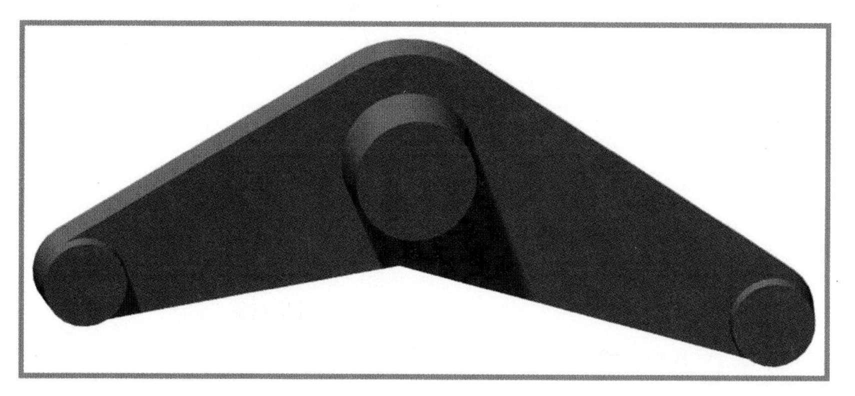

FIGURE 10.4 Projections are added to the machine rocker arm.

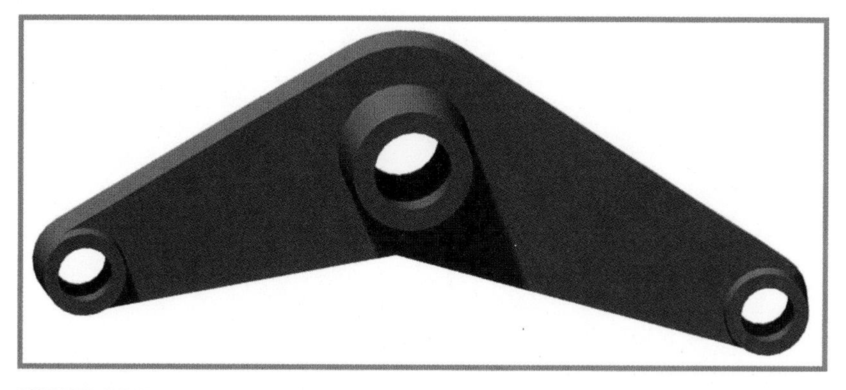

FIGURE 10.5 Holes are subtracted from the machine rocker arm.

After a few more additions and subtractions, we end up with the finished rocker arm that we saw earlier in Figure 10.1.

Stress and thermal flow calculations can be performed on a 3D CAD solid model. Figure 10.6 shows the stresses in our rocker arm.

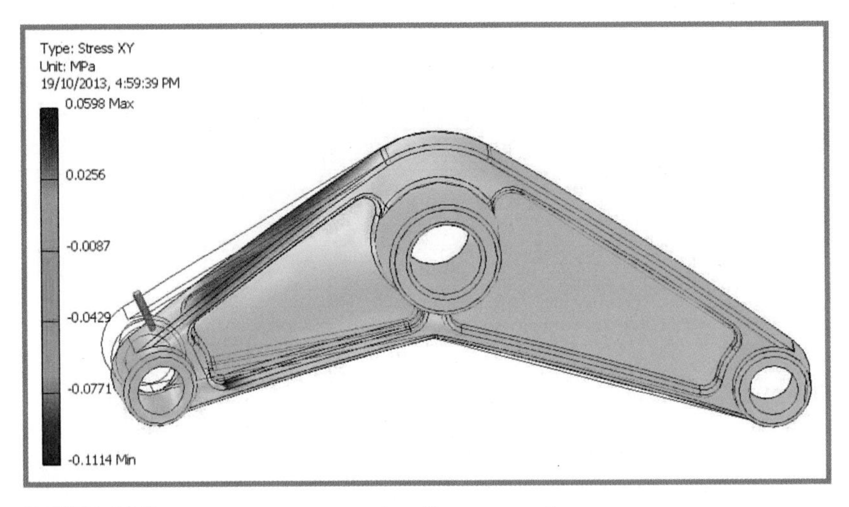

FIGURE 10.6 Our rocker arm is all stressed out.

Just like the real world, two 3D solid CAD models, or even portions of them, can't exist in the same place at the same time. Well, okay, technically they can, but you can perform an interference analysis to show you the offending overlapping regions.

Animated 3D CAD Models

3D CAD models can be animated to show the operation of an assembly, as indicated in Figure 10.7. Okay, trust me. This is set up as an animation in Inventor, but you'll understand the significance of it in a moment.

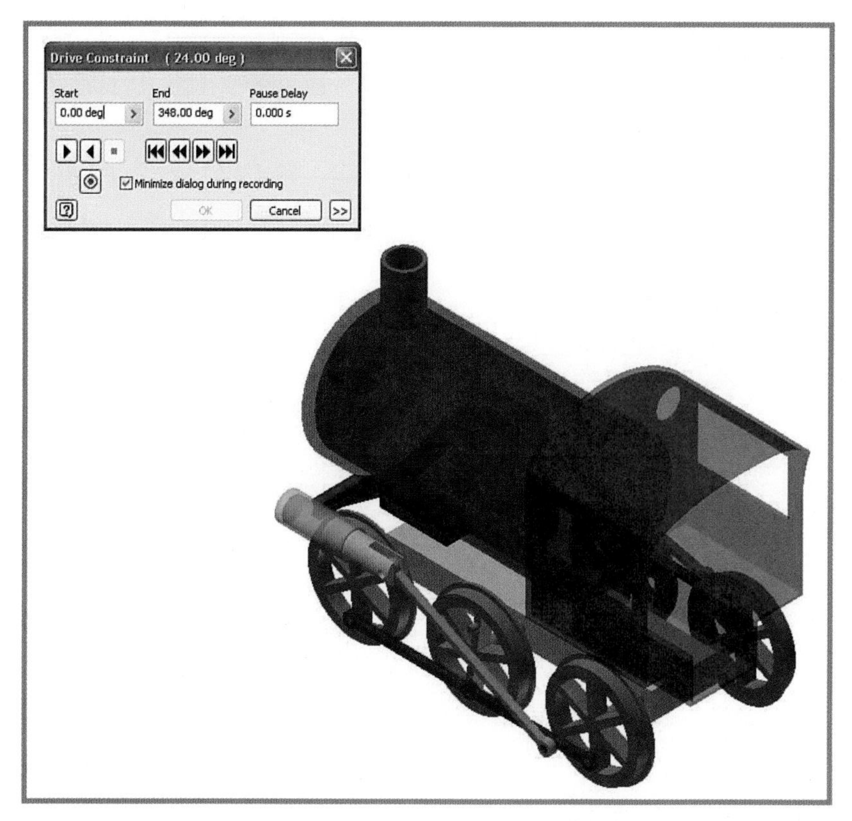

FIGURE 10.7 Our locomotive is ready for a test run.

Oops, the little locomotive that couldn't. When we run the animation shown in Figure 10.7, the wheels turn about three-quarters of a revolution and then everything crashes to a sickening halt. Figure 10.8 shows that the cylinder isn't deep enough and the piston has bottomed out.

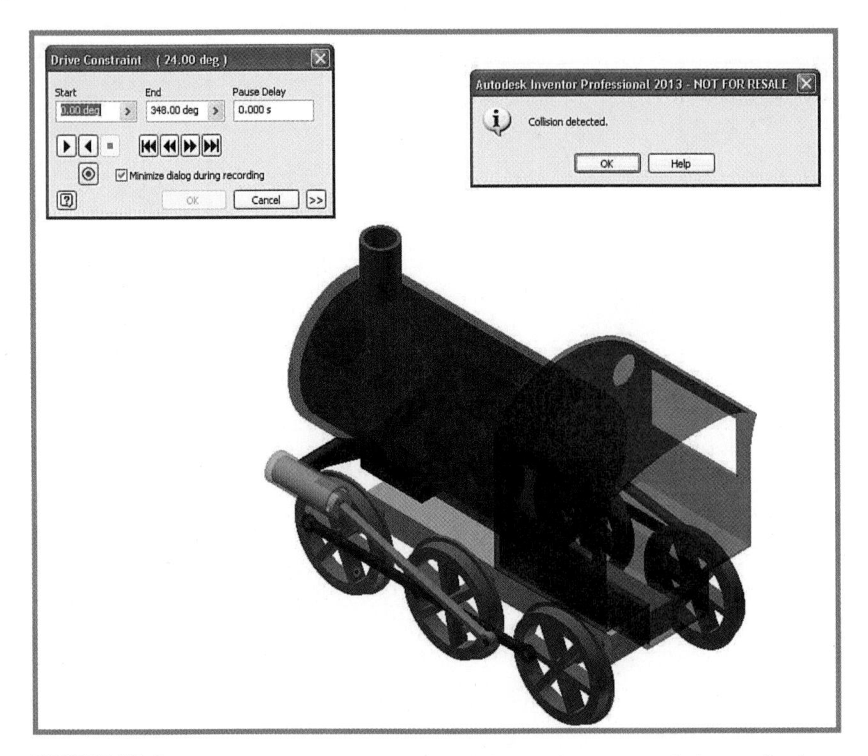

FIGURE 10.8 Crash! The piston has hit the bottom of the cylinder.

Force, velocity, and acceleration analyses can be performed on the animated assembly. If your 3D solid CAD model works onscreen, the real machine probably will, too.

Show Me Some Skin Models

The other type of 3D CAD model comes in two basic variations, and hence the "two and a half kinds" mentioned in the opening paragraph of this chapter. In these types of 3D CAD models, the solid part is represented as an infinitesimally thin outer shell or skin. This type of 3D CAD model is usually called a *surface* model. However, this can be confusing because there are two types of skin—mesh and surface. But a mesh is different from a surface, and there are several types of surface...sometimes you can't tell the players even *with* a program. My brain hurts. I'll refer to them collectively as *skin models*.

Anyway, we'll start with the basic concept. Imagine painting an object and then dissolving the object so you just end up with the paint. That would be a skin model of the object. Similarly, a balloon could be the skin model of a bowling ball.

Figure 10.9 shows the three basic types of 3D CAD models, all of which were produced by revolving a circle about an axis.

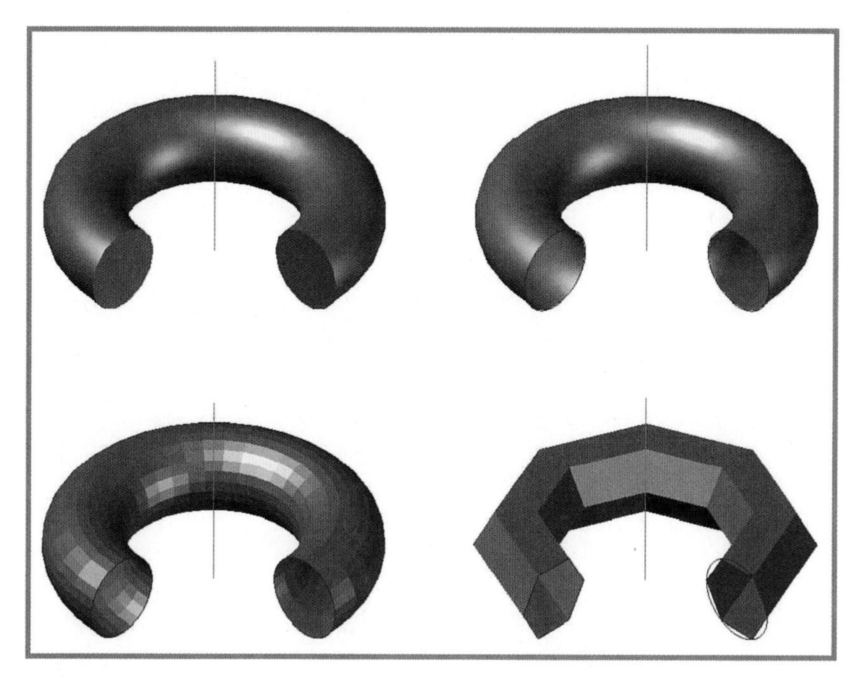

FIGURE 10.9 There are three types of 3D CAD models.

Wait a minute, there are four images in Figure 10.9.

We'll come back to that a little later, but first let's look at the four models starting with Figure 10.9a in the upper-left corner. This is a true 3D solid model. It has full mass properties and is effectively a curved solid bar.

Figure 10.9b in the upper-right corner is a surface model. This is effectively an infinitesimally thin-walled bent tube. It has no mass properties. Surfaces are mathematical representations of curved surfaces, which are created by operations such as extruding or revolving a closed profile, by sweeping a profile along a path, or by lofting between two or more closed profiles. Surfaces flow smoothly, as shown in Figure 10.9b.

NOTE

A complex 3D model can be created by using several surface features. Multiple surfaces can be trimmed where they intersect, and new surfaces can be created to plug openings.

Figure 10.9c in the lower-right corner is a mesh model. A mesh model is composed of a small number of infinitesimally thin, flat plates connected at their edges and vertices in much the same manner as the night club disco mirror balls. It was produced with

AutoCAD's SurfTab1 and SurfTab2 system variables set to their default value of 6. Pretty lumpy doughnut, isn't it? It has no mass properties.

Figure 10.9d in the lower-left corner is also a mesh model. It was produced with AutoCAD's SurfTab1 and SurfTab2 system variables set to 25. It's still a little lumpy, but we could always crank the SurfTab system variables up to their maximum of 32,766.

MAKING A MESHY PART

AutoCAD meshes cannot be Booleaned, but a wide variety of other editing operations are possible. For example, the entire mesh density can be increased or decreased to make the mesh smoother or coarser, and one face of a mesh can be split into two separate mesh elements within the parent mesh. Vertices and edges of mesh elements can be moved to new locations, and the adjacent edges and vertices can move accordingly to maintain the mesh as a single object. Using these techniques, I once saw a person who started from a simple cube and ended up with a mesh model of a commercial jet airliner. Hey, I never said it made any sense, I'm just saying he did it!

By the way, I used a traditional doughnut shape instead of a jelly doughnut because I figured if we cut into the latter, we would get mess properties instead of mass properties.

Skin models are just hollow shells, so you can't perform any of the mass property calculations on them. Similarly, you can't do any interference or collision checking. Two or more skin models can coexist or overlap in the same space without knowing about the others.

I'm sure you've all seen the novelty balloons where a second, smaller balloon is inflated inside a larger transparent one, such as the example shown in Figure 10.10. If these were 3D solid models of a bowling ball and a toy figurine, then interference detection would complain, but because they are skin models we can't even check.

FIGURE 10.10 A skin model of a figurine surviving nicely inside a skin model of a bowling ball.

AutoCAD Versus Inventor

For 3D printing, the final model MUST be airtight with no holes or overhangs, no matter how tiny. This can be a problem when working with surfaces and meshes, but by definition it is not a problem with solids.

This gives Inventor the edge in a great many applications because of its parametric functionality; just change the value of a dimension and everything updates. Inventor works primarily with solids but does have some surface capabilities.

AutoCAD can create and edit surfaces, meshes, and solids. A mesh can be converted to a surface, and meshes and surfaces can be converted to solids if they are contiguous and airtight. You must be in AutoCAD's 3D Modeling workspace to see the ribbon menu that contains the 3D commands.

NOTE

By way of full disclosure, we must also mention that there are other programs besides AutoCAD and Inventor that can produce suitable 3D models, such as SolidWorks from Dassault Systèmes and Solid Edge from Siemens PLM Software.

Summary

As discussed in earlier chapters, you need a 3D digital computer model of an object before you can 3D print it. This chapter discussed surface versus solid models and indicated why solid models are preferable for our needs.

Keep in mind that, historically, surface modelling has had the advantage when creating free-flowing curved surfaces such as car body panels or the casing for a hair dryer, while the prismatic nature of 3D solids tends to make them more suitable for precise shapes such as machine components. More recently, however, the Fusion application from Autodesk allows prismatic solids to be stretched, formed, and manipulated much like surface models. This capability has been added directly within Inventor 2015.

11

Why and How to Use 3D Printing

There are three main uses for 3D printing in the design world. The first is to verify that designs are correct and will work. The second is for form and feel, while a third is for low-volume production.

Let's start with design verification. You may well be asking yourself "Why do I need to 3D print prototypes? I thought modern software can analyze 3D models and can verify that things will work." Hopefully you asked yourself quietly if other people are within listening range.

NOTE

A *prototype* is an early sample or model of an object or a concept that is created to use for testing. When testing of the prototype is successful, other forms are copied or developed from the prototype.

Yes, CAD software can analyze and simulate quite a bit, but as the title of the lead editorial in the Sept. 13, 2007 issue of *Machine Design* (http://machinedesign.com/archive/simulations-are-doomed-succeed) stated, "Simulations are doomed to succeed."

The problem is that we tend to stop analyzing and simulating when we get what we think is a suitable answer, but simulations can be a classic example of garbage in, garbage out (GIGO). If we make the wrong assumptions when setting up the simulation and/or if we feed it incorrect values, then the simulation can give us what appears to be an acceptable result—but it won't match the real world. There is also a strong tendency to think "It came out of a computer, therefore it is correct."

What Can Possibly Go Wrong, Go Wrong, Go Wrong...

Let's start with a couple of quick examples of humans trusting in the results from computer software.

The first example involves an Excel spreadsheet. The student assignment was to create macros within the spreadsheet so it would graph the deflection of a cantilever beam. For most of the students, it was their first encounter with programming, and many were struggling. When their spreadsheets finally stopped complaining about syntax errors, they handed them in. The problem was a common error—omitting one pair of parentheses. The macro would run, but you'd be surprised (and scared) by how many students thought a 4"×8" cantilever beam 12 feet long loaded with 1,000 pounds would deflect 34 million miles. This was a case of incorrect setup.

The next case comes from an exercise in professionally developed course material for a finite element analysis (FEA) session. Following step-by-step instructions, the students set up a vertical steel plate in the CAD software and "welded" an angle bracket to it, as shown in gray in Figure 11.1. They then constrained the bracket to the plate, applied a load to the end of the bracket, performed a stress analysis, and came to the conclusion that the bracket was strong enough.

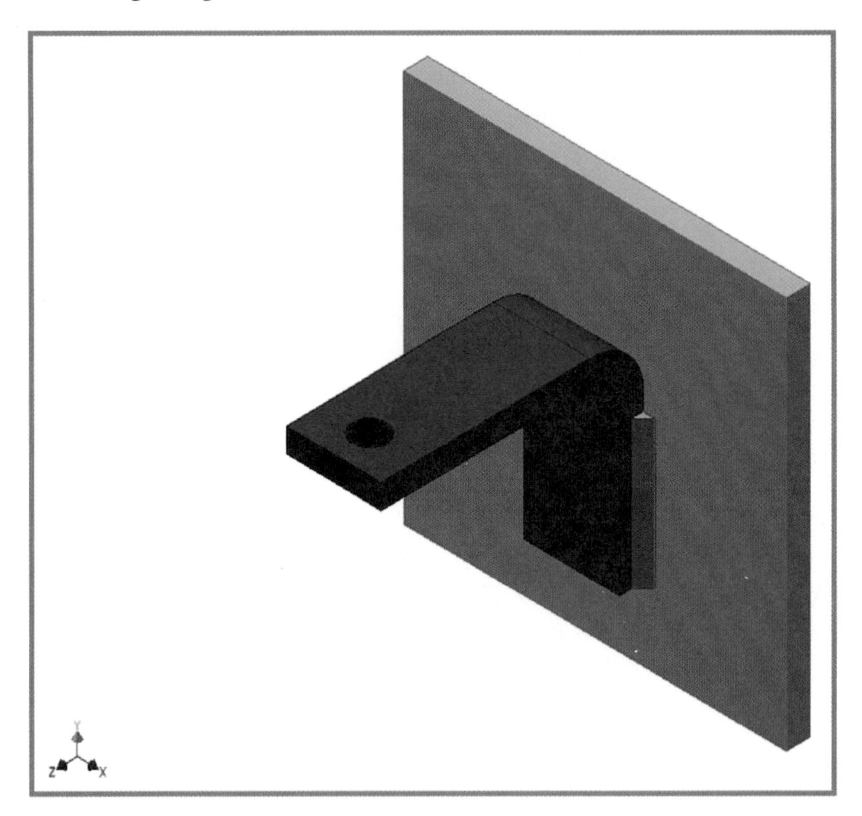

FIGURE 11.1 The basic stress analysis setup.

The problem was that the instructions said to apply a load constraint between the plate and the entire flat face of one leg of the angle as though the bracket was glued in place, as shown in red in Figure 11.2.

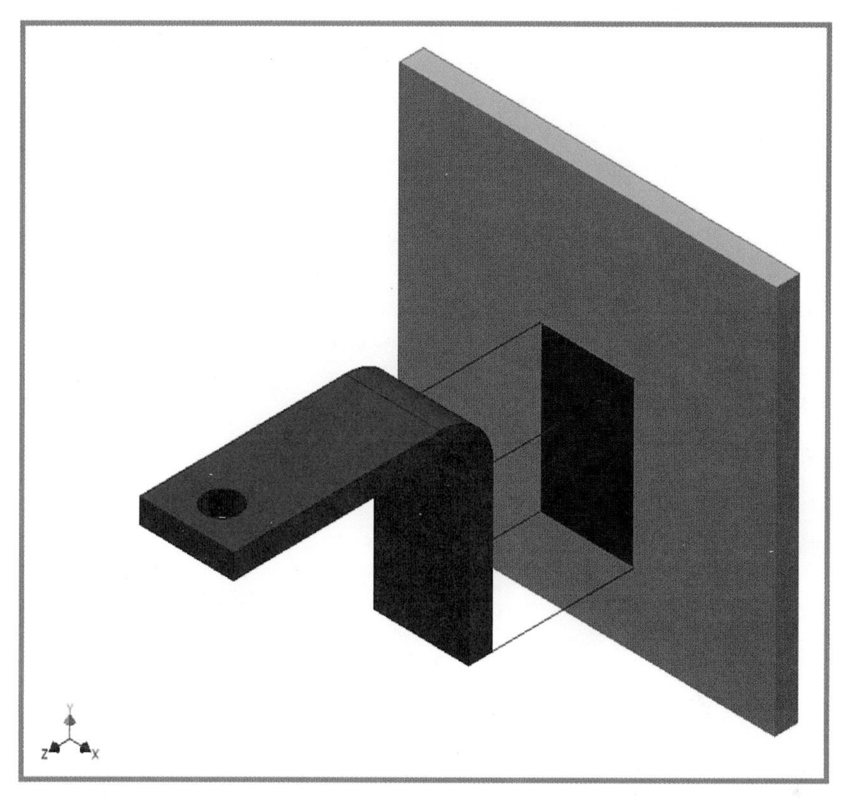

FIGURE 11.2 The stress analysis configuration as per the instructions.

In fact, however, it was attached by two fillet welds, one down each vertical edge of the bracket. Figure 11.3 shows how these welds had only a small fraction of the load-carrying area (shown by the two red strips) compared to the full-face-contact assumption. Therefore, in the real world the bracket would have failed at less than 5% of the specified load. This was a case of incorrect assumptions about how the constraints should be applied.

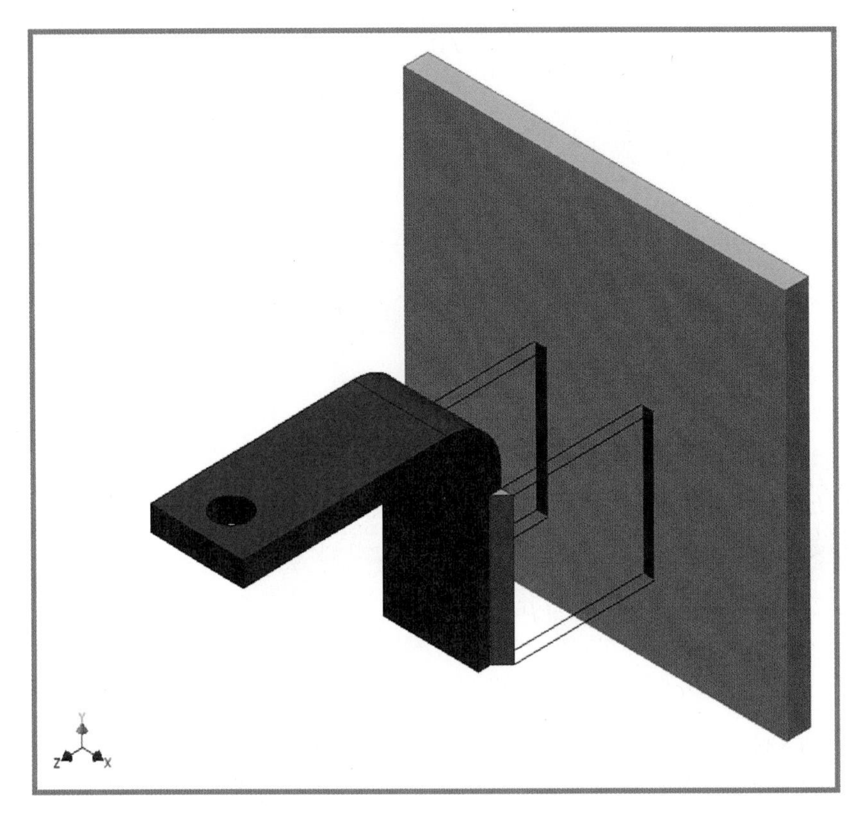

FIGURE 11.3 The stress analysis configuration as per the real world.

Admittedly, these two examples would not have been discovered by using a 3D printed model, but they do point out the hazards of relying too heavily on computer analyses.

Using 3D Prototypes to Verify Designs

Now let's look at two cases where 3D-printed prototypes probably would have helped discover problems.

The first of these two cases involves a very complex 3D CAD model of a Harley chopper custom motorcycle. The model contained every nut, bolt, washer, and even all the balls in each ball bearing. The presenter was very proud of the model and the analyses. Figure 11.4 shows a simplified schematic of the geometry of the rear suspension.

FIGURE 11.4 Keeping the rubber on the road?

You don't need to look too hard to see that the geometry of the rear suspension was such that the rear spring was lifting the wheel up instead of pressing it down onto the road. When the first analysis failed, the presenter had simply reversed the sign of the applied force to get the simulation to succeed. This was a case of incorrect values being used. The bike was already under construction when I pointed out that it wouldn't work, to which the presenter replied "Oh, poop," or words to that effect. A 3D-printed model would have turned up the problem pretty quickly.

The second example is a more generic case that can point out the benefits of 3D-printed models. The problem is that the real world can be different from 3D CAD models in several significant ways. For example, Figure 11.5 shows a CAD model with the hole in the center of the wheel constrained to be concentric with the hole in the center of the bearing.

FIGURE 11.5 What's wrong with this picture?

This looks good onscreen, but Figure 11.6 shows that in the real world if we don't put a shaft between the wheel and the bearing, the wheel will fall off.

FIGURE 11.6 We're shafted because there's no shaft.

Even when we add the shaft to our CAD model, it will still be different from reality. In our CAD model the holes and the shaft remain perfectly concentric, but in the real world everything shifts off-center a bit by the clearances that are required for assembly and running. The red areas in Figure 11.7 indicate this.

A 3D-printed prototype can help determine whether these anomalies are likely to cause problems. Bob Newhart did a great comedy routine about a supermarket manager ordering new grocery carts. He wanted 7 that pulled to the left, 5 that pulled right, and 12 that went "thunka-thunka-thunka".

FIGURE 11.7 In the real world things run off-center because of running clearances.

When relatively simple examples such as these can get messed up, it sort of makes you wonder about the validity of extremely complex analyses such as weather and financial forecasting, doesn't it? Ah, yes, but the answer came out of a computer....

NOTE

3D printing can greatly reduce the cost of prototypes. A few years back the padlock on my boat trailer hitch cost more than $5,000 and took two weeks to produce because it was the first handmade prototype of a product that eventually went into production and sold for about $5.00. Today, a 3D-printed prototype would cost a few dollars and take a few hours to be produced. Granted, it probably would not be strong enough to serve in the real world, but it would help to verify fit and function.

I HAVE A FEELING...

It has been my observation through the years that not everyone thinks like an engineer. Yes, I know, a lot of people would say that this is a good thing, but the bottom line is that very few non-engineers or designers can look at three orthographic views of an object and then visualize the 3D real-world object. Even colored, shaded, perspective views might not be enough.

Here's an extreme example. Our design department invented a radical, totally new product. There was nothing else even similar to it on the market. We hand-built a working prototype. When we showed it to sales, marketing, and company executives, however, we were told it would never sell because the prototype was the wrong color.

A 3D-printed model of a new product can provide a hands-on touch-and-feel experience of a new product before committing to hard tooling or even to an expensive handmade prototype. Are switches and controls in convenient locations? How does it look from different angles and under different lighting conditions? Is it comfortable to hold? I have an electric chain saw that has an annoying ridge right at the base of my thumb when I'm using it. A 3D-printed prototype can provide valuable feedback to the designer. Just make sure it's the right color!

Manufacturing Small Quantities with 3D Printing

3D printing is also known as *additive manufacturing*, and as this terminology implies, it can be used for producing more than just prototypes. It can also be used for manufacturing low-volume quantities, where *low volume* might be 100 parts or less.

The most common forms of 3D printing use thermoplastics such as ABS and PVC, and so might be appropriate for producing small production quantities of plastic parts such as specialized and even personalized novelty items. For example, Autodesk Inventor can create a 3D feature starting from a regular text object, and by changing the content of the text, you can update the model. How about personalized name plates, as shown in Figure 11.8?

FIGURE 11.8 What's in a name?

Creating Metal Parts with 3D Printing

Okay, 3D printing can produce sample parts from plastic or other soft materials, but what if you need metal parts? You might want only one or two for prototype testing before committing a huge pile of money into molds for zinc or aluminum die castings, or perhaps you need only a few dozen production parts for a low-volume product.

No problem. The 3D-printed part can be used as the pattern for sand-cast metal parts. Figure 11.9 shows the basic cope and drag made using the 3D-printed part as the pattern. Now, all the foundry needs to do is to add the sprues, gates, runners, and vents as appropriate, although these could even be added to a derived variant of the basic part so they would already be 3D printed as part of the pattern.

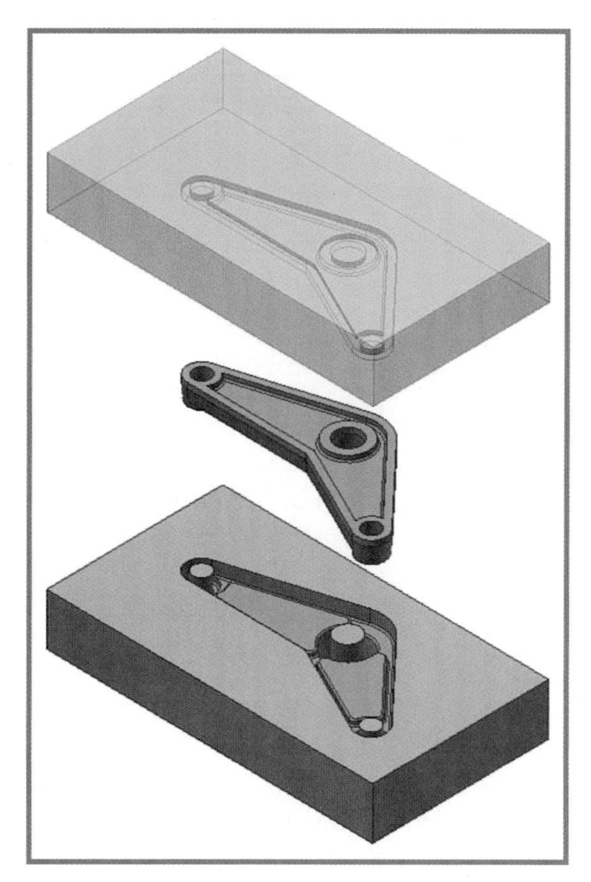

FIGURE 11.9 A plastic 3D print of the rocker arm is used as a pattern to produce the cope and drag for a sand casting in metal.

3D-printed parts can also be used as the pattern for investment casting. Simply dip the part in plaster and then melt or dissolve the pattern as per the standard investment casting process.

Allowing for Shrinkage

One significant point to remember is that almost all materials shrink as they solidify during the casting process, so the part you 3D print to be a casting pattern will have to be larger than the final part by the shrinkage ratio. This can easily be accomplished by using Autodesk Inventor's derived part functionality whereby a new part is created by deriving it from an existing part file.

TIP

Of particular interest in this process is the fact that a scale factor can be set. Aluminum, for example, typically shrinks about 10% depending on the particular alloy and the temperature, so a derived part would use a scale factor of 1.1 to produce a casting pattern that will produce a finished cast part of the correct size.

Using 3D Printing for Large Parts

Speaking of size, 3D printers—especially lower-priced units—typically do not print large parts. There are two workarounds for this problem.

First, a scale model might suffice. I have seen a model of a two-story house scaled down by 1:48 (1/4" = 1' 0", commonly called *quarter-inch scale*). It was a great demonstration model to show prospective buyers before the house was built, and it could help the architect find any anomalies in the design.

Second, larger parts can be printed in smaller sections and then glued together. I have seen a full-size model of a chopper motorcycle (no, not the one referred to earlier; this one was correct) and one of an aircraft jet engine made this way.

Summary

No matter how powerful computer analysis is these days, it's still a very good idea to make a physical sample as a final check. 3D printing can be a fast, economical method for producing such prototypes. It can also be used for producing very small runs of production parts.

Designing Easy-to-Print Parts

For 13 years I taught mechanical engineering at the British Columbia Institute of Technology (BCIT) in Vancouver, Canada. Our program was divided into five options:

- Design
- Manufacturing
- Mechanical systems (heating, ventilating, air conditioning [HVAC])
- Plastics
- Robotics

One of the courses I taught was called "Manufacturing Processes" and covered how things are made. At the start of the term the non-manufacturing students often grumbled about having to take a manufacturing course when they weren't in that option.

The answer is that you can't do a good job of designing something unless you also know how to make it. Certain processes are more appropriate for certain types of parts.

For example, you wouldn't use sheet-metal stampings to make a car engine. Okay, Crosley made cars from 1939 to 1948 that did, but they tended to be noisy and have corrosion problems because of the thin wall sections.

Let's try again. You wouldn't use sand castings to make aluminum car body panels. Yea, I know, Pierce Arrow did for 16 years in the early 1900s and French coachbuilders Hibbard and Darrin used this process in the late 1920s for custom Rolls-Royce bodies, but the labor cost of trying to cast large 1/16"-thick panels proved prohibitive.

Design Versus Make—Know the Process

The examples in the previous section serve to prove my point that designers need to understand manufacturing processes. As the old saying goes, "The devil is in the details." The following section shows how relatively minor design details can make a significant difference to the cost.

Consider the part shown in Figure 12.1. It's an injection-molded plastic part that looks harmless enough.

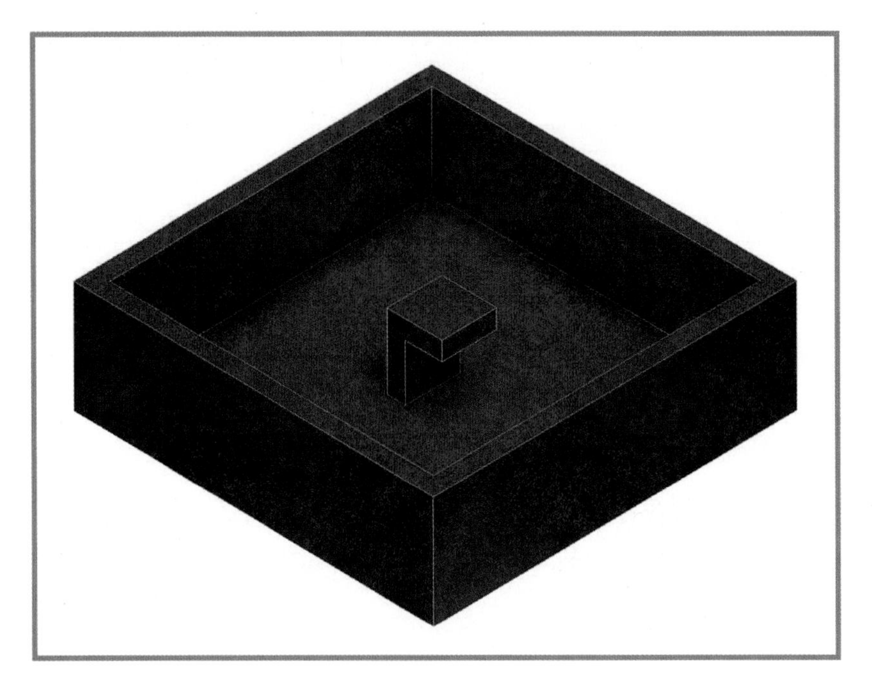

FIGURE 12.1 A simple plastic part.

Okay, so we build the mold, install it in the machine, and shoot in the plastic, as shown in the cross-section of Figure 12.2.

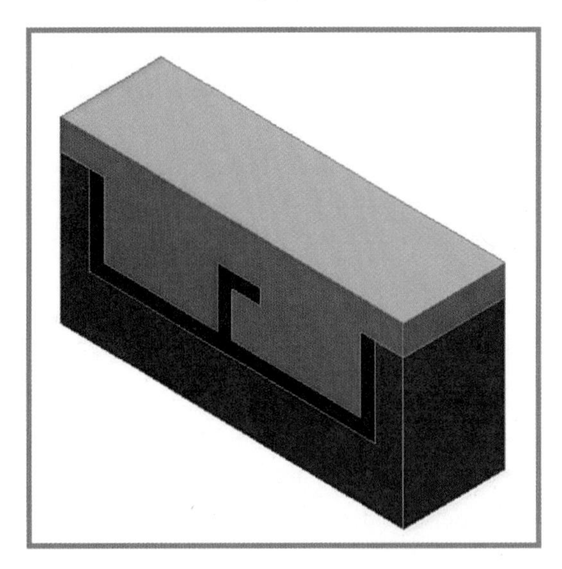

FIGURE 12.2 Our first production run of the new part.

Now it's time to open the mold and remove our part. Oops, Figure 12.3 shows that we have a problem. The red part comes out of the green half of the mold easily enough—or at least it would if I had modelled it with proper draft angles—but the undercut hook on the part won't allow it to be removed from the gray half of the mold.

FIGURE 12.3 Oops, we have a small problem.

We Can Do This the Hard Way...

The expensive way of solving this problem is to build a complex, expensive, delicate, precision mold with a collapsing, retractable core mechanism. The yellow and blue parts in Figure 12.4 show the extra pieces in the mold.

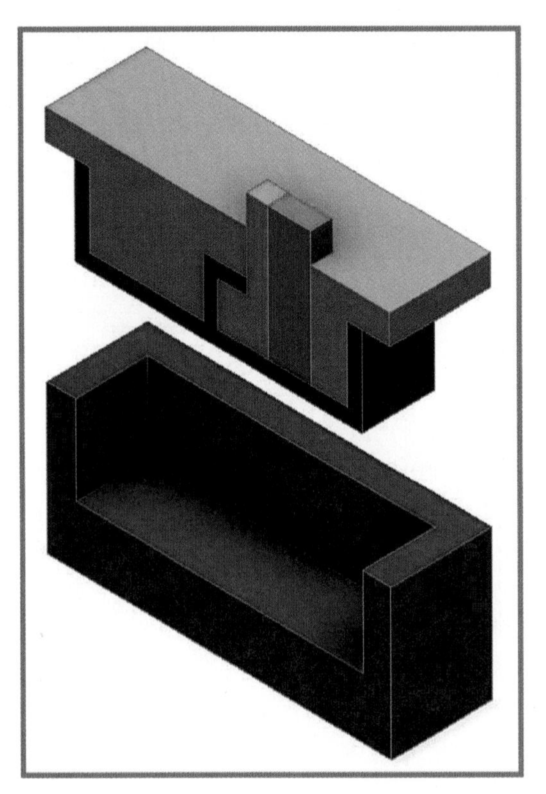

FIGURE 12.4 A complicated multi–sliding-core mold can solve the problem.

Figure 12.5 shows how the blue core has to be pulled up and then the yellow core slid sideways to remove the red part.

FIGURE 12.5 The part can be removed from the mold after the core pieces are shifted.

...Or We Can Do This the Easy Way

The easy way of solving the problem of removing the undercut part from the mold is to add one rectangular hole under the hook, as shown in Figure 12.6.

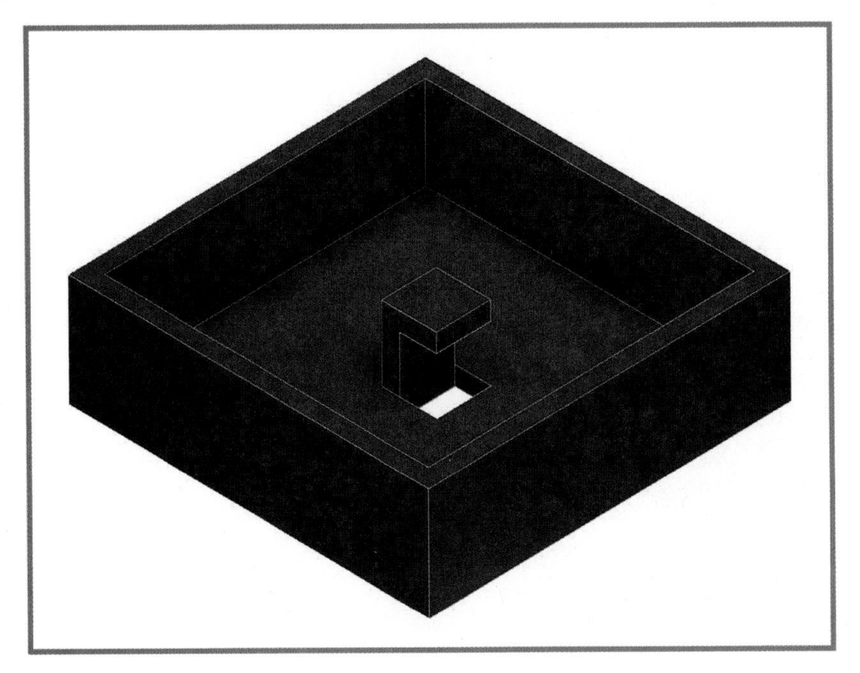

FIGURE 12.6 Adding one small hole solves the problem.

Now the part can be made in a simple two-piece die with no additional cores. Figure 12.7 shows how a finger extends up from the bottom of the green half of the mold to form the rectangular hole and the undercut of the hook.

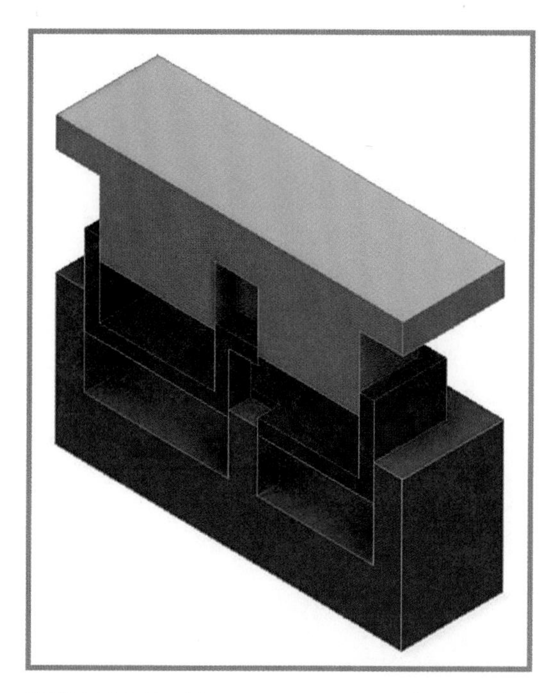

FIGURE 12.7 The part can now be produced using a simple two-piece mold.

The modified part will cost about one-tenth as much as the part that was shown in Figure 12.1. For a good example of this, look at the inside of a typical household smoke detector to see how they formed all the hooks that let the components simply snap in without using any other fasteners.

Helpful Hints to Minimize Problems

So, what does all this have to do with 3D printing? Quite a bit, actually. Unfortunately the current state of the art generally means that 3D printers are not *Star Trek*-like matter replicators that can deliver a real solid object of any old 3D CAD file we throw at them.

Like any other manufacturing process, 3D printers have their strengths, their weaknesses, and their quirks. This is compounded by the fact that although all 3D printers work on the same underlying principle, there are two main process variants, as mentioned in Chapter 2, "Basic Principles of 3D Printing."

The following sections provide a few hints that can help you to minimize problems when 3D printing. Yes, you might have to modify your design a little bit to suit the printer.

Size of the Little Details Matters

One fundamental rule applies to pretty much all printers, however, and that is printer resolution. Chapter 2 explained that all 3D printers work by printing narrow stripes of material one layer at a time.

You won't be able to print anything or any detail small enough that it gets close to the layer thickness or stripe width. You won't make actual-size wristwatch gears on a home hobbyist machine, and in fact probably won't even be able to make them on most high-end commercial machines.

No Visible Means of Support

Look at the simple J-bolt in Figure 12.8. Once again, it appears rather innocuous until it comes time to print it.

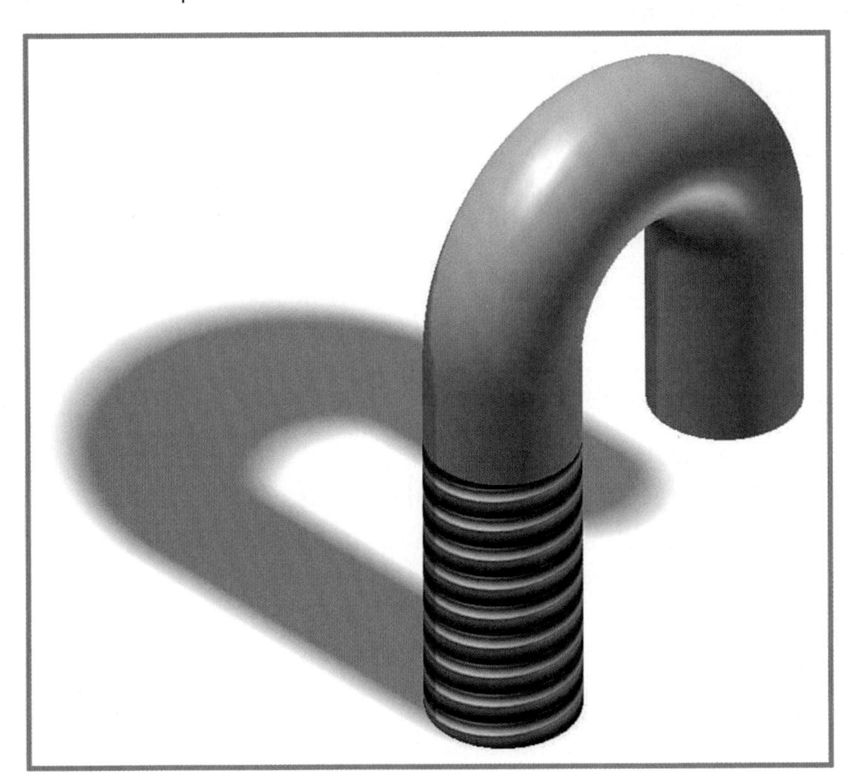

FIGURE 12.8 We want to 3D print a simple J-bolt.

Working up from the bottom, everything is fine until it comes to the bottom of the short leg of the J-bolt. Oops, Figure 12.9 shows that the machine is trying to print in thin air.

FIGURE 12.9 You can't print in thin air.

Gravity is usually fairly reliable, except maybe in weightless conditions in space, so the end bit falls down as shown in Figure 12.10 before the printer has time to get high enough up in the part that the *J* shape bridges across.

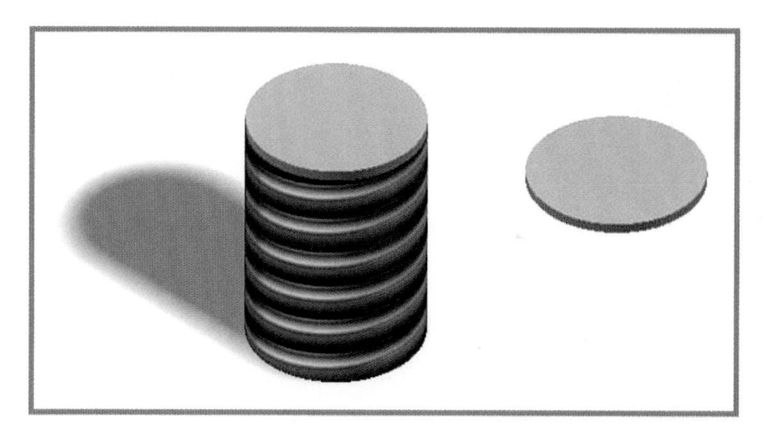

FIGURE 12.10 Ring around the rosie...all fall down!

There are several solutions to this problem.

First, you might not even have a problem. As indicated in Chapter 2 and earlier in this chapter, there are two basic types of 3D printers.

Fusion-Type Printers

The first type uses the fusion principle. In this type, the print space is full of a liquid or a powdered solid and the printer solidifies the liquid or fuses the powdered solid together. The advantage to this type, especially the powdered variant, is that the unused raw material is able to support overhanging bits until it becomes their turn to be bridged across to the rest of the part. Figure 12.11 shows this effect.

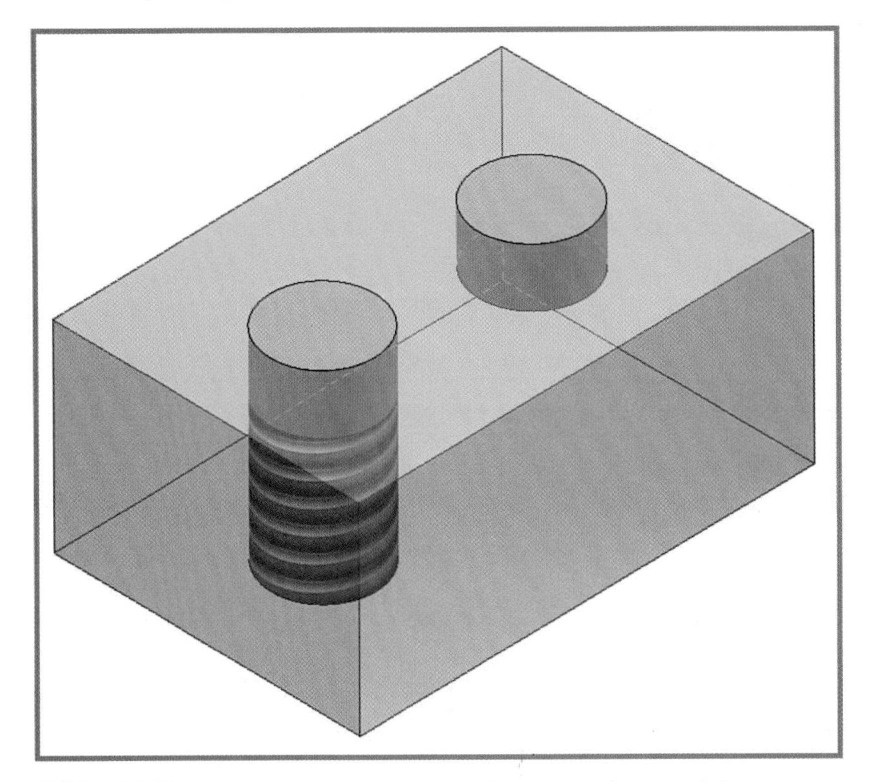

FIGURE 12.11 In fusion-type printers the unused material can support overhanging bits.

A disadvantage of fusion-type 3D printers is that you can't print a sealed hollow shape such as a tennis ball. Well, you can, but you have no way of getting the unused material out of the inside without drilling a hole in the part.

Deposition-Type Printers

The second type uses the deposition or "hot-melt glue gun" principle. This type is subject to the overhanging detail problem. Fortunately, there are several solutions to the problem.

Which Way Is Down?

One solution to the overhanging problem is simply to align the part differently. For example, Figure 12.12 shows the J-bolt lying on its side. If it were 3D printed this way, then nothing would be overhanging.

FIGURE 12.12 Laying the part on its side can overcome the problem of overhanging bits.

Changing the model's orientation isn't always a viable solution for two reasons. One reason is that 3D-printed parts can have a grain direction, much like the grain in wood, due to the layered manner in which they are built. This means that they might be stronger in one direction than another, which can be the wrong direction for a specific application of a given part.

The other reason that alignment might not solve an overhanging problem is that a part may be so complex that no matter which way it faces, there will be an overhanging bit somewhere.

This Is a Hold-Up!

The good news is that if the shape of a part is such that overhanging bits are unavoidable, then you can always add temporary support structures that are removed after the 3D printing process is complete.

The really good news is that you might not have to do anything about them in AutoCAD or Inventor (or any other brand of 3D modeling software, for that matter). When it comes time to do the actual 3D printing, you have to export your model in a file format that 3D printers understand.

Anyway, the 3D printer software inhales this file and translates it into the machine control codes that actually drive the printer. This software lets the operator rotate and align the model as desired. In addition, most brands are able to analyze the model and generate any required temporary support structures.

→ *Detailed information on exporting the file format of your model to a 3D printer is included in Chapter 14, "Exporting Models to a 3D Printer."*

The only slightly bad news is that these temporary support structures must be removed manually. The good news is that higher-end 3D printers are available that can print two or more different materials into one model. The second material can be wax or a water-soluble material that can simply be melted or dissolved to remove the support structures.

Creating Usable 3D-Printable Threads

Did you notice anything unusual in Figures 12.8–12.12? Take a close look at the threads. To reduce the file size, Inventor's Thread function doesn't actually produce anatomically correct 3D threads. It does store the thread specification information in the file, and the presence of threads affects mass properties such as center of gravity, surface area, and weight. However, if you try to 3D print a standard Inventor threaded feature, all you'll get is a smooth cylinder. What looks like threads on the part is just exactly that—Inventor simply paints a flat bitmap onto the cylindrical feature to make it look like a thread.

To create usable 3D-printable threads on a part, you must use Inventor's Coil function. Search for "3D Coil" using Inventor's Help facility to get detailed instructions, but a quick summary follows:

1. Create the cylindrical feature to which you want to apply threads.
2. If necessary, create an appropriate workplane. You can use one of the part's origin planes (X,Y; X,Z; or Y,Z) if they align with the cylindrical feature.
3. Create a 2D sketch of the thread profile on the workplane. This will nominally be a triangle, but real threads have a small radius in the bottom and are truncated a bit at the top.

NOTE

Note that Inventor requires at least a little bit of truncating. For example, a typical Imperial thread is specified as ¼-20 UNC, which means the thread has a pitch (distance from one peak to the next) of 1/20" (0.050"). However, Inventor will complain of a "self-intersecting coil" if your thread profile triangle is wider than 0.49999".

4. On the ribbon, click the **3D Model** tab; then on the **Create** panel click the **Coil** icon.
5. Enter the thread specifications as appropriate. Note that the default is to add the new feature, but you should cut the thread, so click the appropriate button. Figure 12.13 shows a length of threaded rod. The left half shows fake threads created using Inventor's Thread functionality, while the right half consists of real threads created using Inventor's Coil functionality.

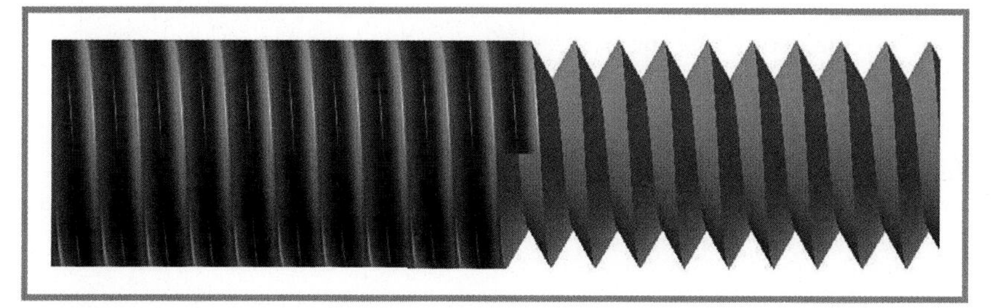

FIGURE 12.13 The left half contains fake threads, while the right half contains real threads.

If you were to 3D print this part, the right half would be threaded but the left half would simply be a smooth cylinder.

If you want to know why Inventor uses fake threads, take a look at the following numbers:

- The bare cylinder has a file size of 75k.
- The cylinder with fake threads has a file size of 85k.
- The cylinder with real threads has a file size of 245k.

Now imagine the snowballing effect of this if you had a model of something like a car engine block with dozens of threaded holes. If they were all real threads, the file size would rapidly hit government-spending numbers. Now create an assembly model of the engine consisting of hundreds of threaded fasteners holding dozens of parts together and processing times; even on a pretty good computer, that can hit geologic time frames.

In addition, fake threads are created considerably faster than real ones.

Solutions to 3D Printing Large Objects

So what do you do if the size of your model is such that it can't be printed using your available printers? No problem. As usual, several solutions are available:

- Print assemblies as individual loose parts, and then assemble them. As you'll see in Chapter 13, "Designing Multipart Models to Print Preassembled," assemblies can often be 3D printed fully assembled, but printing the individual parts and then assembling them is usually the best approach anyway.
- You can 3D print large individual parts in sections. You then glue them together.

- Service bureaus are available that have larger machines. You upload your file to the bureau's website and then they 3D print it and ship it to you.

→ *Go to Chapter 15, "Using Inventor to Print Directly to Third-Party 3D Printing Services," for more information on service bureaus.*

There are two ways to 3D print a house:

- Architects and designers often like to have a small scale model for demonstration purposes. The scaling is usually done by the printer software so you don't need to worry about it in your CAD software.
- You can always go full size. That's right; 3D printers that fit on a flat-deck semi-trailer now exist. They can print a 1,200-square-foot concrete house in about a day. Similarly, 3D printers exist that can create full-size synthetic sandstone decorative panels for architectural detailing.

Conversely, the printer software can scale up small things such as watch gears to build a demonstration model that is much larger. This can require a bit of adjustment when exporting the 3D print file from your CAD software, though.

→ *Go to Chapter 14, "Exporting Models to a 3D Printer," to learn more about exporting 3D print files.*

Summary

This chapter explained the importance of knowing how manufacturing processes work to ensure that parts can be manufactured economically. It then described some of the unique characteristics of the 3D printing process. The tips and tricks explained in this chapter will help ensure that parts can be 3D printed easily.

The next chapter goes on to discuss assemblies. Unlike most other manufacturing processes, 3D printers can often produce several different parts at one time, and the parts come out of the printer already assembled. As you will see in the next chapter, however, this isn't always a good thing.

Designing Multipart Models to Print Preassembled

An interesting feature of 3D printing that always attracts attention at demonstrations is the fact that it is possible to print more than one part at a time, fully assembled, all in one hit. To make this work, you need to be aware of a few details that we cover in this chapter. On the other hand, we also explain why this might not be a good idea under certain circumstances.

Effects of Printer Resolution on Parts

When 3D printing is a complete assembly that involves parts that can move relative to each other, the first significant point to note is printer resolution. As discussed in Chapter 12, "Designing Easy-to-Print Parts," the layered nature of 3D printing means you can't print small details that approach printer resolution. When it comes to assemblies, this problem typically applies to the running clearance between moving parts. Usually, the clearances need to be made larger.

For example, a typical sleeve or journal bearing might have a running clearance of about 0.003" (0.08 mm), but the thinnest layer your printer can produce might be 0.010" (0.25 mm). If you try to print your CAD model as is, you will probably find that instead of an assembly of moving parts you have one big lump. All, or at least many, of the moving parts will have fused together.

The bad news is that you'll have to adjust those clearances manually in your CAD model before you export the printer file.

TIP

Don't forget to set the clearances back to normal before you order the regular production parts, or you can follow the procedure I outline a little later in this chapter.

This includes any surfaces you want to be able to move relative to each other, such as:

- Shafts in sleeve or journal bearings.
- Linear sliding surfaces.
- Cams and cam followers.

- Ball bearings (more on this later in this chapter).
- Threads, which need to be specifically cut—not Inventor's bitmaps, as discussed in Chapter 12. You can also cut or tap the threads manually after the part has been 3D printed. This includes mechanisms such as screw jacks and linear actuators as well as threaded fasteners.

Using Derived Part Functionality

Usually, the best way of handling the bearing clearance modifications in Inventor is to use its derived part functionality. This is actually quite a simple process:

1. Launch Inventor, and start a new standard part.

> **TIP**
>
> If your default setting of Inventor launches a new part with a sketch already active, close the sketch. You can even delete it if you want to.

2. Click the **Manage** tab of Inventor's ribbon.
3. Click **Derive** in the **Insert** panel. This opens the **File Open** dialog box shown in Figure 13.1.

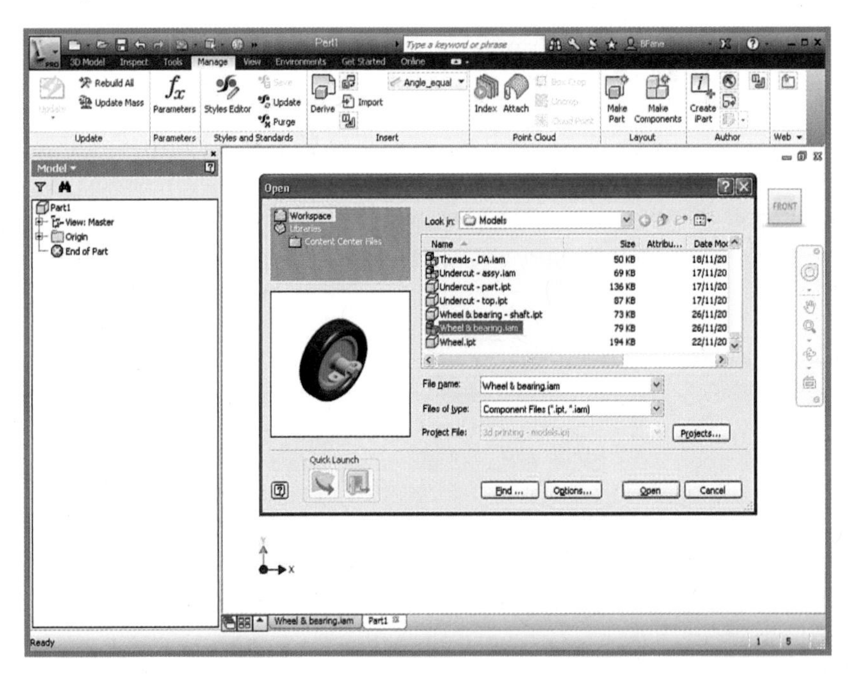

FIGURE 13.1 Selecting the assembly to be the parent of a derived part.

4. Browse to and open the desired parent assembly. The **Derived Assembly** dialog box opens, as shown in Figure 13.2.
5. This dialog box has a number of options under four different tabs, but for our purposes right now, we'll simply take all the defaults. Click **OK**.

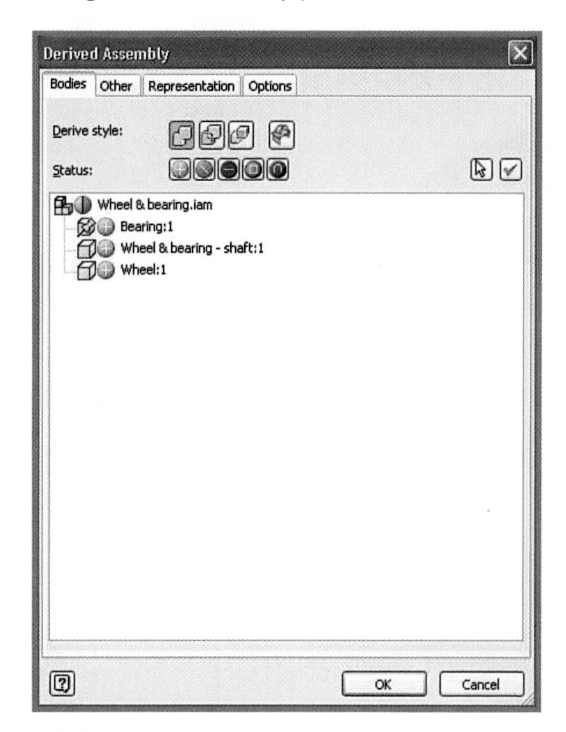

FIGURE 13.2 Selecting the options for deriving a new part from an assembly.

The result looks exactly like the original assembly, but with one major difference. It is now a single solid body with only one feature per component part, which means the individual features cannot be edited other than to suppress or un-suppress them. Each feature of the new part carries forward the color of its source part.

Ah, but now comes the fun part. The new part itself can be edited like any other part. For example, we could do the following:

1. Create a 2D sketch plane on the end of the shaft.
2. Project the OD of the shaft onto the sketch plane if your default settings don't do it automatically.
3. Draw a concentric circle that's a little smaller than the shaft.
4. Dimension the difference between the two diameters by simply starting the **Dimension** function and then selecting the two circles.
5. Finish the sketch.

6. Select to **Extrude Cut** the circle through **All**.

Bingo! Figure 13.3 shows how we have increased the clearance between the shaft and the bearing, and between the shaft and the wheel, without affecting the original parts or assembly at all. This newly derived and modified part will 3D print just like a single part but will come out of the printer as three assembled pieces with running clearance between them.

FIGURE 13.3 We now have running clearance between the shaft and the other two parts.

Better yet, any changes made to the original parts will reflect through to the derived part automatically. Feature sizes in Inventor can be associatively linked to features in other parts, so that if we change the shaft diameter, then the bore size of the bearing and the wheel will also change. Our derived part will update accordingly while still maintaining the extra clearance required for 3D printing.

TIP

In step 6 of the previous example, you can also extrude to the front face of the wheel so the shaft is still locked to the wheel but loose in the bearing.

Resolving Interference Problems

Merrily we roll along...or maybe not. Interference between parts can be problematic. So why would we want interference? Usually interference between parts is a no-no, but sometimes we deliberately want it. For example, bearings might be press-fit into a slightly undersize housing, or a gear might be press-fit onto a slightly oversize shaft to prevent loose play and wobble.

An issue that can arise here is that the software for some printers might or might not be confused when presented with parts that overlap, such as interference fits. If your intention is to have the real-world parts locked together, the safest approach is to use Inventor's Derived functionality as described earlier in this chapter.

Refer to Figure 13.2 and note the yellow circle with a + (plus) sign beside each part. Clicking a yellow circle scrolls it through several options, including a red circle with a - (minus) sign, which says to subtract (or in our case, ignore) the selected part. We can thus derive a single part consisting of just the wheel and shaft.

More than one part can be derived from a single assembly, and a new part can be derived from a single part, so in our example we could derive the wheel and shaft as a single part from the assembly and then derive a new single part from just the bearing. The derived bearing would then be edited to increase its bore.

Problems Unique to AutoCAD

AutoCAD has similar functionality, but for a different reason. In Inventor, you start building a part by creating a suitable sketch profile and then turning it into a solid. Subsequent features are then sketched onto and Booleaned (added to, subtracted from, or intersected with) the base feature.

AutoCAD, on the other hand, makes you create each feature separately. The Boolean operations to create a complex part are then performed separately on each feature. For our purposes this isn't a problem, and in fact is exactly what we want. Simply add the wheel and shaft parts together to get a single part, and then cut a cylinder from the shaft or bearing to get the required clearance.

Issues with Ball and Roller Bearings

Ball and roller bearings are another special case. In the real world, the internal parts are built with interference fits. In a ball bearing, for example, the balls are slightly larger than the grooves in the races, so there is no loose play or sloppiness. The problem is that when you use Inventor's Design Accelerator functionality to insert a ball bearing into an assembly, it creates an anatomically correct model, as shown in Figure 13.4.

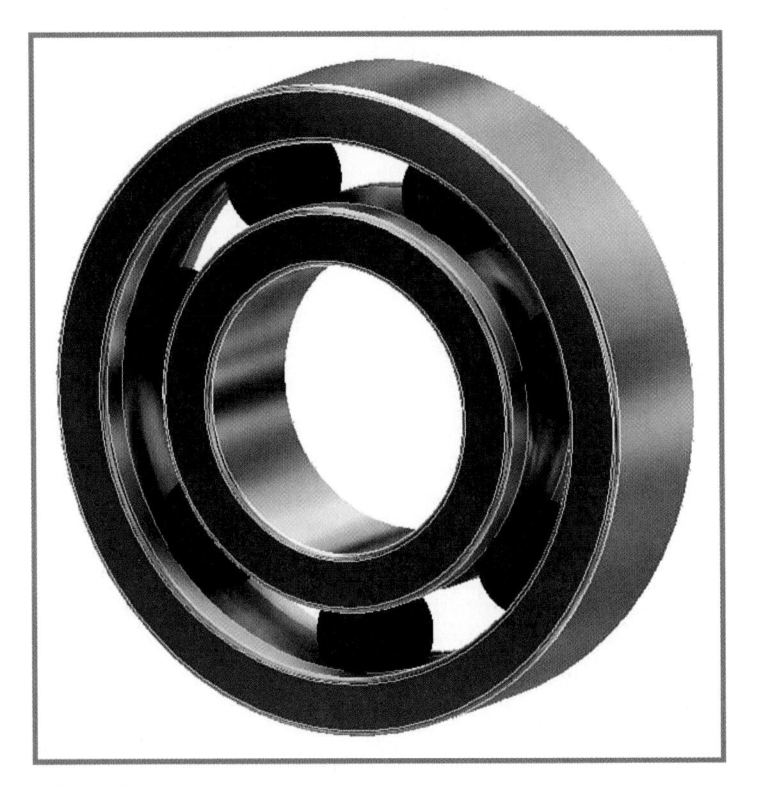

FIGURE 13.4 Inventor's Design Accelerator functionality produces anatomically correct bearings.

The problem is that it is indeed anatomically correct, including the interference fit. The cutaway section view of Figure 13.5 shows how everything fuses together into one lump when you 3D print it.

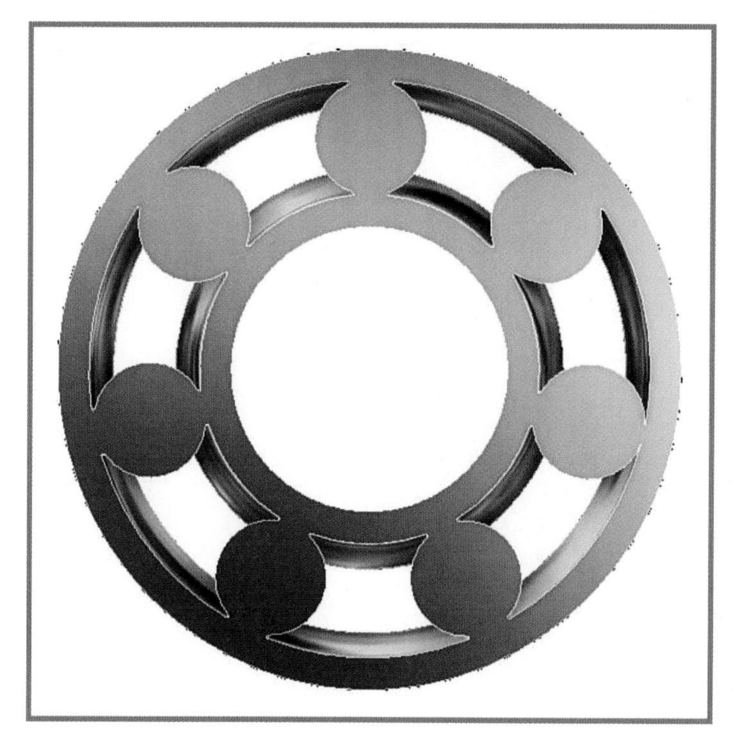

FIGURE 13.5 Oops, not much of an antifriction bearing, is it?

Unfortunately you can't edit the Design Accelerator bearings because they are built to international standards and it normally wouldn't make sense to allow users to edit a standard item.

Ball and Bearing Solutions

There are two solutions. One is to laboriously model the bearing yourself, making the balls slightly undersize so they 3D print with a bit of clearance. The resulting 3D printed bearing shown in Figure 13.6 will be a little loose and sloppy, but it will rotate.

FIGURE 13.6 This bearing works, but it's a little loose and a lot of work.

The other possibility is to recognize that there are really only three user-critical dimensions to a ball bearing—the OD, the ID, and the width.

Figure 13.7 shows a cut-away model of a simple replacement for a ball bearing. It is a single part containing just a single feature revolved from one sketch, but it can be printed as a single part. It might not operate as smoothly as a real-life metal ball bearing, but it will rotate in a 3D-printed assembly.

FIGURE 13.7 Simple replacement for a 3D-printed ball bearing.

NOTE

Inventor enables you to easily substitute parts within an assembly and to have alternative configurations using different parts, so you can easily switch your model back and forth between a real bearing and your printable one. If you use a lot of bearings in different sizes, you also should learn about Inventor's iPart Factories. They let you instantly generate a new part with different dimensions based on a single factory part.

Considerations Before Using 3D Printing for Parts

Just because something can be done, doesn't mean it should or must be done. 3D printing a set of parts preassembled introduces a few more things to consider:

- Too much clearance between bearing surfaces can cause misalignments and jams. The mechanism probably won't work as smoothly as the final production version produced by "normal" processes.
- Plain bushing bearings and linear sliding surfaces won't have smooth, polished finishes. If parts are printed separately, then they can be sanded and polished before assembly. Printing parts separately usually eliminates the need to modify the CAD model to produce the required running clearances.
- Because they aren't touching, moving parts will probably need temporary support structures. The 3D printer software can often add these automatically, but they need to be manually removed from the final printed assembly.
- Overhangs can exist in an assembly that wouldn't if the parts were printed separately. With the parts being printed in place, their positions cannot be oriented for optimal 3D printing. Again, temporary support structures may be needed.
- Most 3D printing involves plastics or corn starch. 3D-printed springs usually don't work all that well.

You can print things that can't be made any other way. This could be good or bad. It's fun doing gimmick demos for trade shows or as advertising pieces.

Variants of the gear assembly shown in Figure 13.8 are popular. Rotating any one gray part causes them all to rotate, but the unit cannot be dismantled. The bad news is that 3D printing can be dangerous as prototypes for other (especially metal) manufacturing processes. Check out the next section of this chapter, which provides more details.

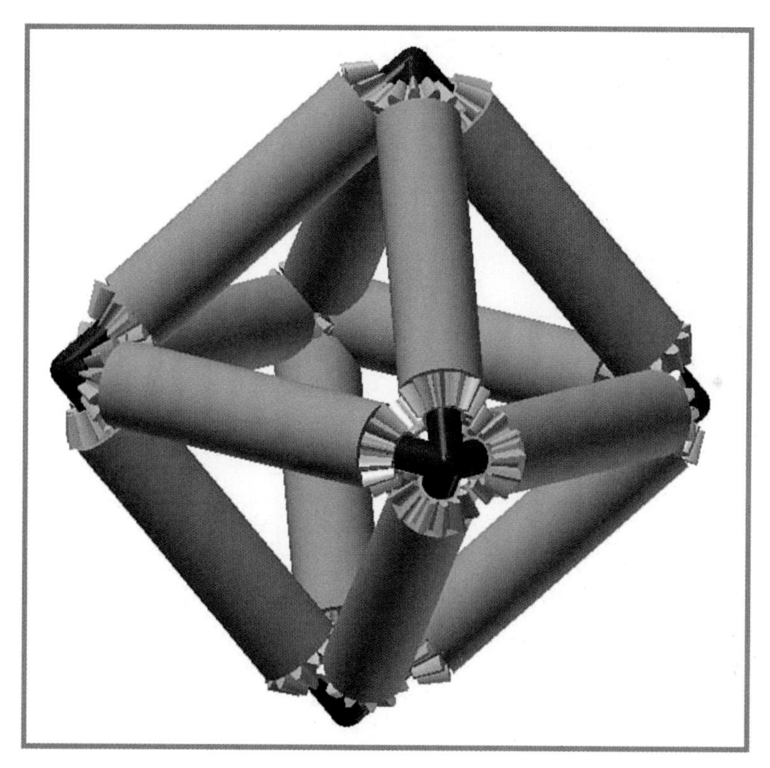

FIGURE 13.8 This can be 3D printed as an assembly. Rotating any one shaft rotates them all.

When Not to Use 3D Printing

Here's a scenario that illustrates why you might not want to 3D print a preassembled, multipart model.

Let's say that you're a designer for the Whatt Manufacturing Company. It has been decided to convert your company's most popular product, the Whatt Shamacallit, into a portable model. Your assignment is to design the new wheels for it.

You're aware of the KISS (Keep It Simple, Stupid) principle of design, so you come up with the very efficient design shown in Figure 13.9.

FIGURE 13.9 This is a simple, efficient design.

You perform a kinematic analysis and a finite-element stress analysis on your CAD model. You 3D print a preassembled prototype out of plastic, and check it out for size and fit on a sample of the current product. Everything looks good, so you release the design for production.

When the initial production order of 10,000 steel parts arrives, you get a panic call from the assembly shop floor. Figure 13.10 shows why.

FIGURE 13.10 Now all we need to do is to assemble the parts.

Oops, it's impossible to assemble the two metal parts. Oh, poop, or words to that effect.

Summary

The bottom line is that if you are 3D printing an assembly to be used as a prototype for other manufacturing processes, then you are usually better off 3D printing the individual parts and then assembling them manually. The primary exception to this rule might be if you know that certain subassemblies are going to be permanently attached together before going into a final assembly. This could include things like purchased subassemblies, weldments, interference fits, and parts held together with permanent adhesives. This is becoming less of an issue as printers working directly in metal get better, but only for low-volume items where it is financially viable to 3D print the entire production run.

The next chapter discusses how to export your models for 3D printing, and fortunately there is a setting that lets you export each component part in an assembly to its own individual printer file all at one time.

14

Exporting Models to a 3D Printer

Exporting your 3D CAD model to a 3D printer can be very simple, or it can be slightly less simple. There are two things we need to settle straight away.

The first thing to note is that at the time of writing this chapter the only way to get your CAD model to a 3D printer is by exporting it in a special file format. Nearly all 3D printers accept the same file format, which is known as an STL (STereoLithography) file.

The second thing to note is that Autodesk is strongly rumored to be working on direct support for 3D printers; Microsoft has already done this in Windows 8.1. Even if this is done, you might not be running the latest release of their software for a while, which means the comments in this chapter still apply.

Let's start with the very simple case.

Exporting STL Files

The instructions for AutoCAD and Inventor are almost identical and will probably be very similar for most other 3D CAD packages. In fact, if you can perform a Save as to a different file format from within almost any computer program, then you are well on your way to being able to export a CAD file to a 3D printer.

Using Inventor to Export Files

Let's start by seeing how to export an STL file for a single part from within Inventor:

1. Make sure your model is up-to-date by checking the **Local Update** button near the upper-left corner of the Inventor screen, as shown in Figure 14.1. Gray is okay, but if it's yellow, click it to invoke the update.

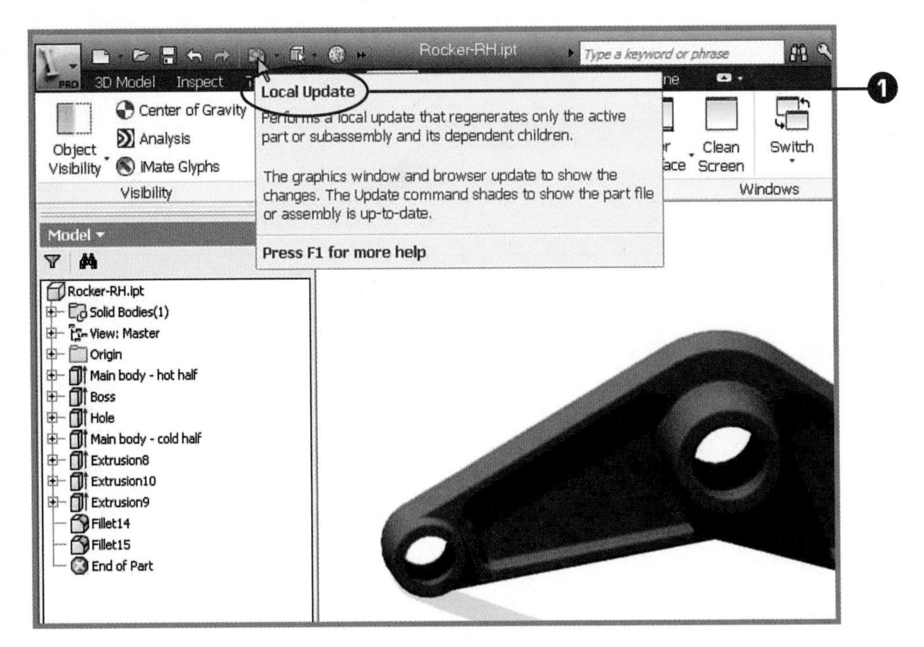

FIGURE 14.1 Make sure your model is up-to-date before exporting it to an STL file.

The next three steps in the process are shown in Figure 14.2.

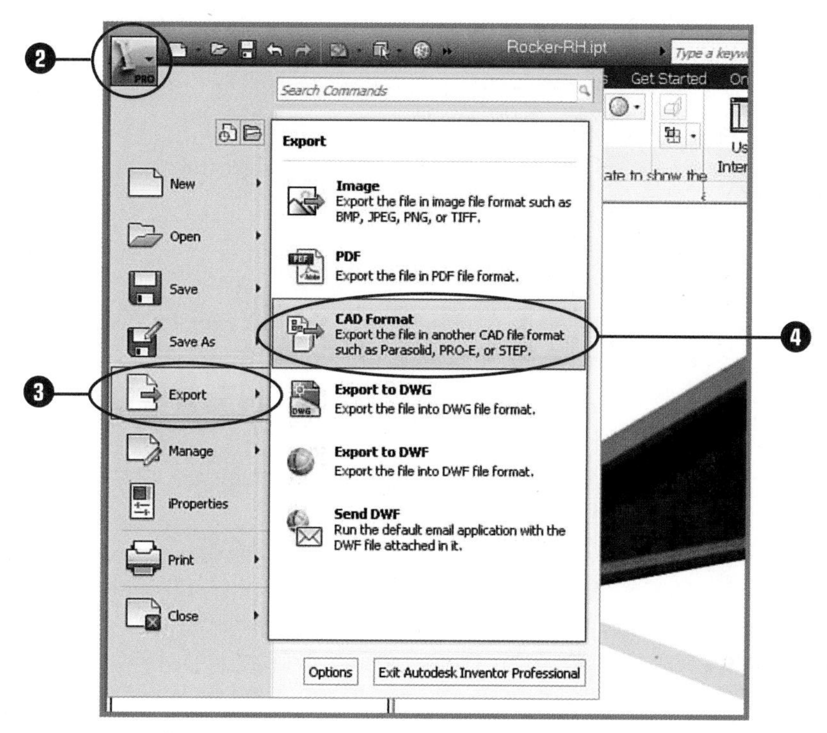

FIGURE 14.2 Here are the next three steps in the export process.

2. Click the **Application** menu (the big yellow I) at the upper left of the screen.
3. Click **Export**.
4. Click **CAD Format**.

   The final three steps should be fairly obvious to anyone who has used a Windows file dialog box.

5. Click the **Save as Type** window in the file dialog box.
6. Scroll down and click **STL files (.stl)**.
7. Click **Save**.

All done! That was easy, wasn't it? Ah, but there is a potential issue to be covered in the "Inventor 2013 Users, Do NOT Try This At Home" section later in this chapter.

Using AutoCAD to Export Files

Now here are the steps for exporting an STL file of a 3D model from AutoCAD:

1. Click the **Application** menu (the big red *A*) at the upper left of the screen.
2. Click **Export**.
3. Click **Other Formats**.

4. Click the **Save as Type** window in the file dialog box.
5. Scroll down and click **Lithography (*.stl)**.
6. Click **Save**.

AutoCAD prompts you to export select solids or watertight meshes. Select the 3D model or models you want, and then press **Enter**.

All done! That was easy, wasn't it?

So far so good, but there are still a few issues to be resolved. The good news is that you might not be the person who has to resolve them because all of the following can usually be handled by the 3D printer's software after it has inhaled your STL file.

Scale

Inventor's Export function is not limited to creating an STL file that depicts the CAD model at its full size, but it can also scale it up or down.

The following section gives four possible scenarios as to when you might want to do this. AutoCAD doesn't support scaling of STL files, but the following comments still apply. AutoCAD users have to rely on the 3D printer's software to handle setting the scale factor:

- **Size**—You probably don't want to print a full-size 3D model of a house, and anyway you can't—unless you have a 3D printer the size of a semi-trailer truck (which do exist, by the way, and can print a house). More typically, an architect or a designer will want a desktop-sized display model, scaled accordingly.
- **Scale**—You might want to print with a different scale on different axes. Topographical maps, for example, are often 3D printed with the vertical scale exaggerated by 10x.
- **Shrinkage**—This is a particular factor when using a 3D printer of the hot-melt-deposition variety. The plastic contracts as it cools and solidifies. This is a function of the particular plastic and of the operating temperature. This also applies if you will be using the 3D printed part as a pattern for a sand casting. The sand casting will contract as it cools, so the pattern needs to be made a little oversize to compensate. Sometimes two wrongs do make a right.
- **Units**—You might have modeled in Imperial units, but your printer works in metric units, or vice versa.

NOTE

As you consider the orientation of your model, refer to other chapters in this book that address the issue of overhang in certain types of 3D printers, including the need for possible temporary support structures and how orientation of the part can influence these factors. The software provided with the 3D printer can reorient the part to minimize this issue.

TIP

If you are using AutoCAD, this is about as far as you need to go, but you might be interested in the "Something Completely Nerdy" section later in this chapter.

Optional Extras

As you have seen in the preceding section, it is very simple for both AutoCAD and Inventor users to export a basic STL file that can be used by the software for most 3D printers.

In the following section you'll see how Inventor users, on the other hand, can continue with a slightly less-simple option:

1. Start by following steps 1–6 of the export procedure outlined in the "Using Inventor to Export Files" earlier in this book.
2. Before completing step 7, however, notice the Options button that appears at the bottom of the file dialog box. Click it to bring up the STL File Save As Options dialog box, as shown in Figure 14.3, and note the following choices:

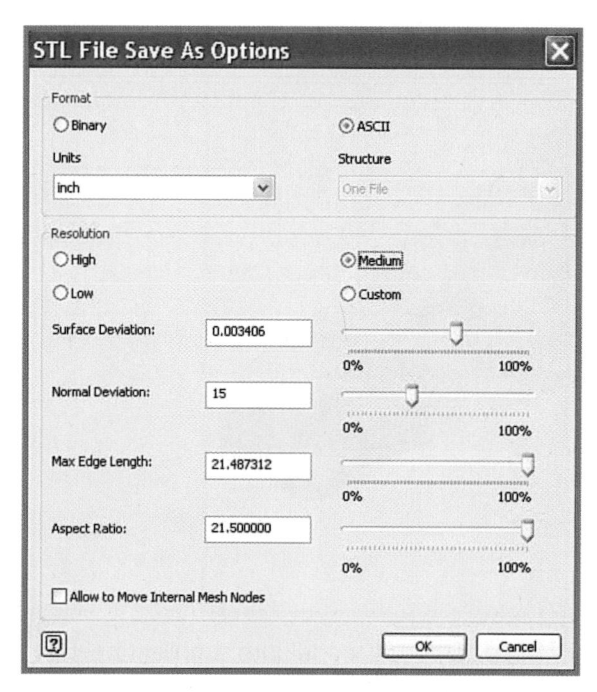

FIGURE 14.3 Inventor adds several options to the export process.

a. **ASCII versus Binary**—This determines the format for the STL file. The "Something Completely Nerdy" section later in this chapter explains exactly what is in an STL file. The only difference between these two formats is that when saved as an ASCII file, it is a simple plain-text format that you can read using Notepad, while saved as a binary file, it is a compressed format. Binary files are much smaller and can be opened by most 3D printer software, but if your printer software can't read a binary file, try ASCII.

b. **Units**—Inventor can rescale your model from inches to millimeters or vice versa as it creates the STL file.

c. **Resolution**—An STL file converts your 3D CAD model into a huge pile of triangular faces that approximate the flat and curved surfaces in your model, as shown in Figure 14.4. More information about the Resolution option is included in the next section.

d. **Allow to Move Internal Mesh Nodes**—I'm not exactly sure what this does or how it works, but the Inventor Help facility simply states that it can be useful when modeling spiral curves.

FIGURE 14.4 An STL file converts your nice, smooth model into a collection of flat, triangular facets.

Resolution Setting Options

Inventor lets you choose between Low, Medium, High, or Custom resolutions to define the size of the triangles. It defaults to Medium, which is usually adequate for most parts going to a typical 3D printer and is approximately the same as the fixed output from AutoCAD.

TIP

Setting the resolution much higher than the 3D printer's capabilities doesn't produce smoother results, but it can produce huge file sizes. 3D printer software can usually dumb down a higher-resolution file. If at some point you want to print on a higher-resolution printer, go a little high if you are in doubt.

The main reason for changing the resolution is if the actual size of your 3D printed part is significantly different from the size of the CAD model. This works both ways.

For example, you might be 3D printing a wristwatch gear at 20 times life size as a trade show demonstration model. If you export at the default Medium resolution and then scale it up, you'll end up with a pretty coarse model that will probably be missing a few teeth. You'll need to export it at a higher resolution so it will scale properly.

Conversely, you might be 3D printing the desktop-size presentation model of an architect's high-rise office tower. The 3D-printed model doesn't need to, and won't even be able to, show the screwdriver slots in the cabinet door hinge screws.

You can play with Custom settings. The farther to the left you move the sliders, the smaller the triangles will be; however, the number of them will increase and hence the file size will increase exponentially.

Assemblies as Separate Files Option

If you are exporting an assembly model, then there will be an additional check box labelled Assemblies as Separate Files. I have some good news and some bad news.

The good news is that as I indicated in Chapter 13, "Designing Multipart Models to Print Preassembled," Inventor can export all the parts in an assembly as individual part files, all in one hit. The bad news is that it exports *every* instance of *every* part into separate files.

Let's say you have an assembly model of a piece of machinery. This machine is held together by 100 identical threaded faster sets, each fastener set consisting of 1 bolt, 2 flat washers, 1 lock washer, and 1 nut. When you export the STL files as separate parts, you'll get 100 identical bolt files, 200 identical flat washer files, 100 identical lock washer files, and 100 identical nut files—for a grand total of 500 files plus all the other parts in the machine.

The good news is that Inventor has several methods available to work around this problem, all based on the fact that it has several capabilities for simplifying assemblies. I won't go into all the gory details on how to do this, but the following notes will give you a general idea of to how to proceed:

- Any instances of component parts that you suppress in the browser tree won't export.
- You can tell Inventor to ignore all component library items. In particular, this includes threaded fasteners, which are usually the biggest problem.
- You can use Inventor's shrink wrap functions to simplify the assembly by merging some parts together and ignoring others to produce a simpler assembly.
- You can use Inventor's derived part functions to merge and/or ignore components in an assembly. As you've seen, it's easy to produce STL files, and so before long you'll have quite a collection of them, to the extent that you can easily forget what the STL part for a given file looks like. You might even have received a file from someone else and want to know what it looks like. The next section shows you how to do this.

Viewing STL Parts

If you want to see what the STL part looks like, Inventor 2013 and later are able to open STL files. The procedure is simple: you just select **STL Files** from the **Files of Type** drop-down list in the regular **File Open** dialog box and then select the file you want.

If you want to see the individual facets more clearly, as shown earlier in Figure 14.4, right-click the last item in the browser, just above the **End of Part** item, and select **Show Mesh Edges**.

CAUTION

Inventor 2013 users MUST read the following section before attempting this.

Inventor 2013 Users, Do NOT Try This at Home

For that matter, don't try this at the office either until you have read this entire section. Oh, it starts off innocently enough, but as you will soon see, it takes an ugly turn very quickly.

The good news is that Inventor 2013 is able to open an STL file so you can see what the exported model looks like. The bad news is...well, read on:

1. Create a complex Inventor part (.ipt) file that includes a number of features and parametric relationships.
2. Save the part file using an appropriate name.
3. Export it as an .stl file using the same name as the .ipt file. This is the default that Inventor 2013 offers, and therein lies the danger.
4. Close the .ipt file.

5. Open the .stl file.
6. Close the .stl file without saving it.
7. Open the .ipt file again. Horrors! Here's the bad news! It now consists of one single dumb mesh object that can't be edited because it lost the original solid model, its individual features, and all its parametrics! The not-quite-so-catastrophic news is that with Inventor 2013 you can retrieve it from its Old Versions folder as long as you don't save the new one-lump version first.
8. To avoid this problem, here are three iron-clad rules to which you must rigidly adhere:
   a. NEVER accept the default filename when exporting an STL file. You MUST give it a name different from any part or assembly model that exists or is ever likely to exist in the same folder. I would suggest adding something like **STL-** to the front of every exported STL file, but you will soon find a bunch of near-duplicate names turning up. For example, Part-1.IPT would export as STL-Part-1.**STL**, and if you ever open it in Inventor, it will produce STL-Part-1.**IPT**, which can subsequently be deleted.
   b. Always obey rule number 1.
   c. Never break rule number 2.

Inventor 2014's Own File Naming Quirk

Inventor 2014 and later versions do not have this problem, but they do have a bizarre twist of their own. You can't edit the .stl mesh, but you can create new sketches and features that have absolutely no connection to the .stl mesh. Even when you do a minor action such as zooming, rotating, or panning the view, when you close the file, Inventor 2014 displays the dialog box shown in Figure 14.5 to ask if you want to save the file.

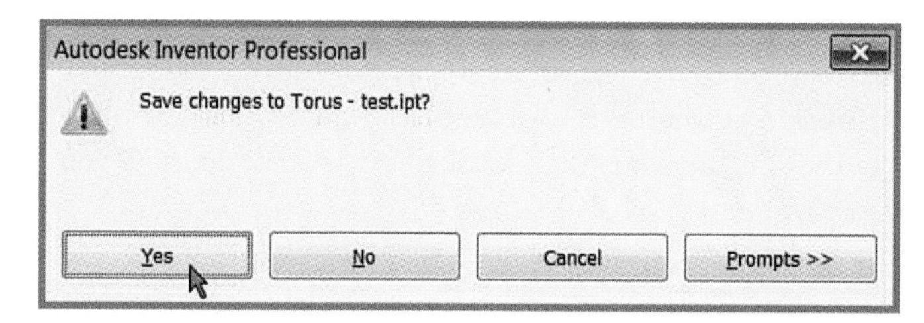

FIGURE 14.5 Check the fine print in the Inventor 2014 dialog box.

The interesting bit here is the fine print in the dialog box. Notice something odd? Wait a minute! You opened the .stl file but it's asking if you want to save the .ipt file! If you click *"Yes"* then Inventor asks for a file name but objects if you try to use the current .stl name, objecting because the .ipt file already exists. Strange.

Using STL Files to Translate CAD Models

The STL file format is not limited to 3D printing. It can also be used to translate models from one CAD system to another, with the two main limitations discussed here. As indicated previously in this chapter, Inventor 2013 and later can open an STL file. The problem is that it is simply a dumb mesh that can't be edited.

If you are an Inventor 2014 or later user and are an Autodesk Subscription member, you can go to the Autodesk Labs website and download an experimental add-in that lets you turn an .stl mesh into an Inventor solid body. It obviously won't have any of the underlying sketches and parametrics, but at least it can be edited by adding, subtracting, or intersecting with newer features. Autodesk Labs apps are often previews of what might turn up in later releases.

The other limitation is that curved surfaces end up being approximated by a series of flat facets, so some smoothness is lost.

Something Completely Nerdy

Okay, so let's face it, you're pretty much a nerd if you got this far into this book to begin with, so we might as well add the icing to the cake. If you don't like to publicly admit that you're a nerd, then you can skip the rest of this chapter because the information in it isn't essential to producing viable 3D-printed objects. However, if you're a computer geek like me, it can be fun to know what's happening inside the magic process.

Getting from your computer to a 3D printer is a two-file process. Earlier in this chapter I discussed exporting an STL file from your CAD system. So what happens next? The 3D printer software inhales it and translates it into a different file type, which is then used to actually control the 3D printer.

As part of this process, the software can do things like rotate the alignment to eliminate or reduce overhangs, add temporary support structures if necessary, and scale the model up a tiny bit to allow for the shrinkage that occurs when the plastic cools and solidifies—if you are sending the model to that type of printer.

Okay, so your 3D printer software has inhaled and processed your STL file. Now let's take a look at what it sends to your 3D printer.

The STL File Format Explained

The STL (STeroLithography) file format is an industry standard recognized by pretty much any 3D printer's software. It defines a surface model that represents the outer skin of the 3D solid model as a series of flat triangles.

Wait a minute! In Chapter 10, "The Difference Between Surface and Solid Models," I spoke against using surface models for 3D modelling work. Among other shortcomings, it is easy to build parts that can't exist and the editing of surface models can often produce leaks.

So why does the 3D printing industry use surface models? Let me count the whys:

- The solid modelling engines of the major CAD packages are different and proprietary. Translation from them can be difficult.
- The file format of 3D CAD modelers typically changes with every release and is usually not backward-compatible; for example, a 2013 file cannot be opened by 2012 software.
- Problems with surface models typically arise when models are edited, but it's relatively easy to produce an airtight surface model when it is all done in one hit while the file is exported.
- The .stl file format is extremely simple, as you can see in Listing 14.1, and it has not changed over time.

LISTING 14.1 STL File Format Sample

```
facet normal -9.921285e-001 -4.726090e-002 1.159632e-001
    outer loop
      vertex    4.000000e+000 4.898587e-016 1.224257e-015
      vertex    4.026955e+000 -4.931598e-016 2.306159e-001
      vertex    3.981888e+000 3.802242e-001 9.771099e-016
    endloop
  endfacet
  facet normal -9.831436e-001 -1.413547e-001 -1.159632e-001
    outer loop
      vertex    4.008721e+000 3.827864e-001 -2.306159e-001
      vertex    3.981888e+000 3.802242e-001 9.771099e-016
      vertex    3.954183e+000 7.621063e-001 -2.306159e-001
    endloop
  endfacet
  facet normal -9.133251e-001 -2.215703e-001 -3.416780e-001
    outer loop
      vertex    4.032160e+000 7.771351e-001 -4.487992e-001
      vertex    3.954183e+000 7.621063e-001 -2.306159e-001
      vertex    3.940031e+000 1.156897e+000 -4.487992e-001
    endloop
  endfacet
```

Okay, it isn't really as complicated as it first looks. Let's try this again, but in Listing 14.2 I converted the numbers from scientific format to general number format using Excel.

LISTING 14.2 STL File Format Using General Number Format

```
facet normal    -0.9921    -4.7261    0.1160
  outer loop
    vertex      4.0000     0.0000     0.0000
    vertex      4.0270     0.0000     0.2306
    vertex      3.9819     0.3802     0.0000
  endloop
endfacet
facet normal    -0.9831    -1.4135    -0.1160
  outer loop
    vertex      4.0087     0.3828     -0.2306
    vertex      3.9819     0.3802     0.0000
    vertex      3.9542     0.7621     -0.2306
  endloop
endfacet
facet normal    -0.9133    -2.2157    -0.3417
  outer loop
    vertex      4.0322     0.7771     -0.4488
    vertex      3.9542     0.7621     -0.2306
    vertex      3.9400     1.1569     -0.4488
  endloop
endfacet
```

The Three Sides of STL Files

This snippet in Listing 14.1 and the alternative version shown in Listing 14.2 from an STL file define three triangles. Each triangle is a facet on the surface of the model.

As you can easily see, it starts with a `facet normal` line. Three points define a triangular plane that has a top and bottom (or in and out). The `facet normal` is a direction vector to tell the software which side of the triangle faces out.

The three `vertex` lines simply contain the X, Y, and Z coordinates of the three vertices of the triangle, while the `outer loop`, `endloop`, and `endfacet` lines are simply delimeters for the data for each triangle.

And that's it. The only requirements are that each vertex of each triangle must have one vertex of at least two other triangles whose coordinates match exactly and that there must be exactly enough triangles to enclose an airtight volume. The simplest possible STL file would be four triangles that define a triangular pyramid, while the STL file for a cone could have hundreds of triangles with a common vertex at the apex of the cone.

NOTE

It's theoretically possible to create an .stl file from scratch or to edit an existing one just using Notepad. The requirements of the previous paragraph, however, make this effectively impossible for anything much more complex than a cube.

G-Code Used to Send STL Files

Many years ago the machining industry began using numerical control for machine tools. This meant that instead of an operator standing there turning a wheel by hand to make the cutting tool move, the machine used a stepper motor to turn the wheel. When the stepper motor received an electrical pulse, it moved exactly one step. The cutting tool could be moved a precise distance by sending it an exact number of pulses. This was known as *numerical control*, and machines originally used paper tape with punched holes to generate the electrical pulses.

As computers came along, they were adapted to generate the control pulses, so this became known as *computer numerical control (CNC)*. A standard universal programming language called G-Code was developed. It is used to control lathes, milling machines, punch presses, routers, 3D printers, and more.

Wait a minute! 3D printers? Yes. Virtually all 3D printers use G-Code to control the print head. Why not? If G-Code can control the position of the cutter in a milling machine, then it can control the print head of a 3D printer. It's an industry-standard programming language that rarely changes, at least not at the relatively simple level required by 3D printers.

The following is an excerpt from the G-Code used to print our rocker part on a MakerBot 3D printer:

```
G1 X38.106 Y104.382 Z0.1 F1350.0 E2.8808
G1 X39.347 Y108.027 Z0.1 F1350.0 E3.0217
G1 X42.047 Y111.497 Z0.1 F1350.0 E3.1825
G1 X44.062 Y112.969 Z0.1 F1350.0 E3.2738
```

It's actually very easy to understand:

- `G1` means "move the print head from where it is now to the following coordinate triplet."
- `X`, `Y`, and `Z` are obviously the values for the coordinate triplet.
- `F1350.0` defines the speed ("feed") at which to move the print head.
- `E`*n*`.`*nnn* defines how much extrudate material to feed to the print head during the move. Note that this isn't necessarily the same as the straight-line distance between the points and may be higher or lower depending on whether the printer wants to produce a thicker or thinner line of material. This code will not usually be present on a fusion type of printer.

A significant point to note is that although STL and G-Code files use X, Y, and Z coordinates, most of the time the coordinates aren't the same. For example, a flat, rectangular surface on the bottom of a part requires only three triangles and hence six coordinate triplets in an STL file, whereas the 3D printer may have to take hundreds of passes to cover the same area and therefore its G-Code file will be much larger.

Another point to note is that the 3D printer doesn't necessarily follow the nice neat parallel-row pattern described in Chapter 2, "Basic Principles of 3D Printing." Instead its software optimizes the print head movements to get the best results, as shown in Figure 14.6.

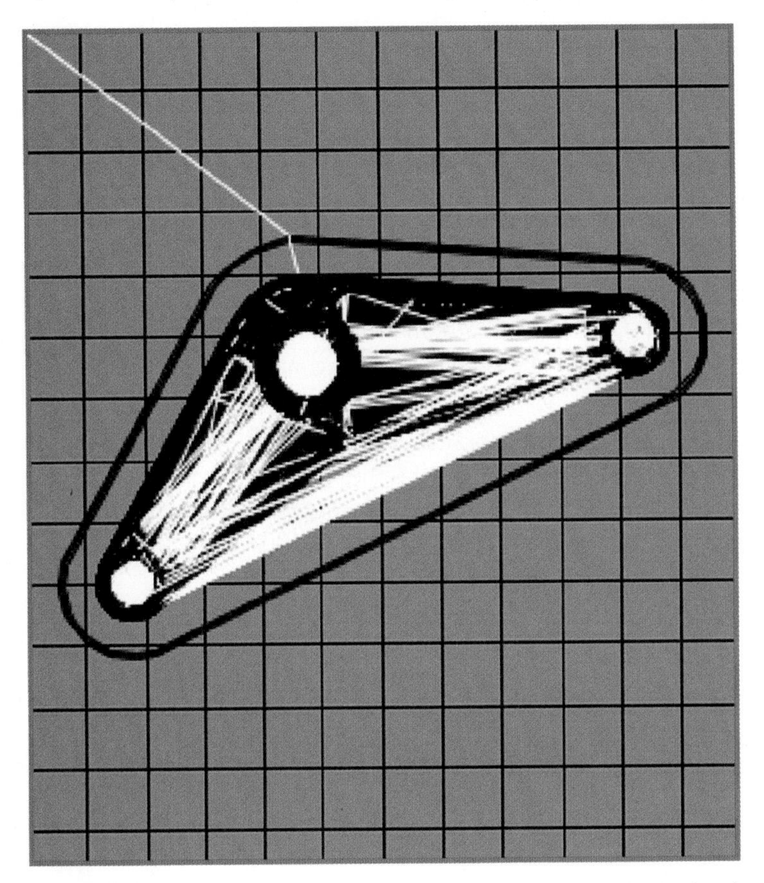

FIGURE 14.6 The 3D printer software optimizes the print head movements.

Summary

This chapter has shown you how easy it is to export your 3D CAD model in a format that can be sent to your 3D printer.

This is good as far as it goes, but what if you don't own a printer that is suitable for your current needs, or what if you don't even own one at all? No problem. The next chapter shows you how to send your STL file to an outside 3D printer service bureau directly from within Inventor.

Using Inventor to Print Directly to Third-Party 3D Printing Services

Okay, so you have finished modeling the "ultimate" design and now you want to 3D print one for executive, sales, and customer presentation purposes. Oops, you have one or more small problems:

- Your 3D printer isn't big enough to handle the model.
- You want a higher-resolution, better-quality print than your 3D printer can produce.
- You want to use a material that your 3D printer doesn't support.
- You don't even own a 3D printer.

This leads us to Life Rule #1: "Never bring up a problem unless you already have a solution." In the present case, a solution is a 3D print service that owns one or more printers with a range of capabilities. There are several nationwide service bureaus that you can use—there may even be one near you.

After you select a 3D print service, you send them the STL file you exported from AutoCAD or Inventor, pay them, and they courier your 3D printout to you. It's that simple.

To make things even simpler, Inventor 2013 and later let you connect to your choice of one of several service bureaus directly from within Inventor. You can play with option settings such as resolution and material, obtain a price quote, order, and pay—all from within Inventor.

The next section shows how this works.

Connecting to a Third-Party Print Service

Connecting Inventor to a 3D print service is easy, as you'll see in the following section. The following example show how this works:

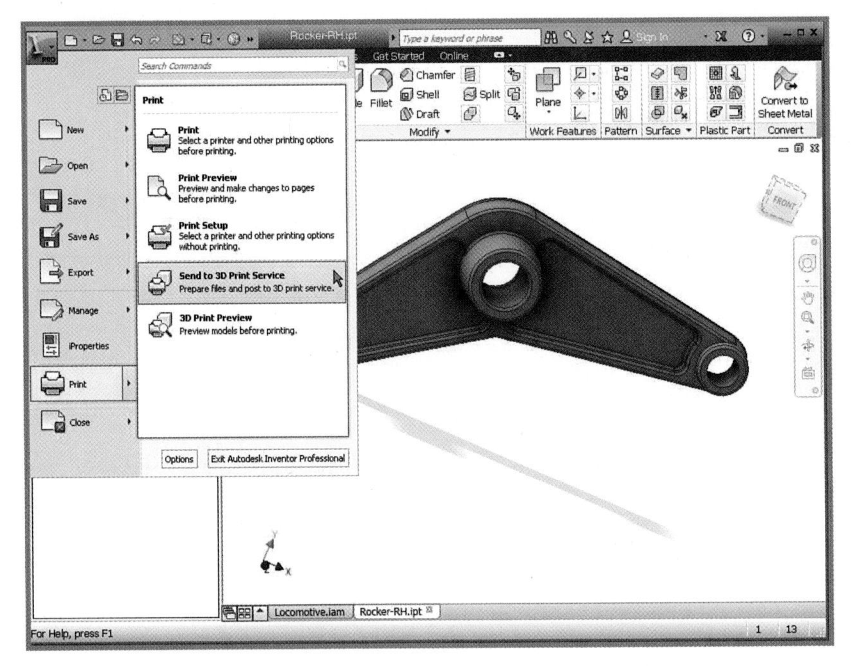

FIGURE 15.1 Starting up Inventor's Send to 3D Print Service function.

1. Open an Inventor part file.
2. Click the **Application** menu (the big yellow *I*) at the top left of the screen.
3. Hover your cursor over or click the **Print** button.
4. Click Send to 3D Print Service.
5. Click **OK** in the **Send to 3D Print Service** dialog box that appears (see Figure 15.1).
6. We now come to the slightly tricky part. Inventor wants you to save the STL file it needs to produce to send to the print service, so it displays the Save Copy As file dialog box and defaults to using the base name of the source part or assembly.

CAUTION

I warned you about this in Chapter 14, "Exporting Models to a 3D Printer," but I'll warn you again—do **NOT** accept the default. Instead, modify it. In Chapter 14 I suggested adding something like **-STL** to the filename. The problem is that if you use the default name and anyone ever opens the STL file in Inventor 2013, then the source .ipt or .iam file will get overwritten without warning, converting it to a dumb 3D mesh. This happens even if the files came from Inventor 2014 or later and even if they close the .stl file without saving it.

7. Inventor now shifts you to an Autodesk webpage, as shown in Figure 15.2. It invites you to **Select a Service Provider to Print Your Inventor 3D Model**. It also provides a brief description of their services and capabilities. Figure 15.2 shows only the top portion of the Autodesk webpage. In the real world you can scroll down through a listing of providers.

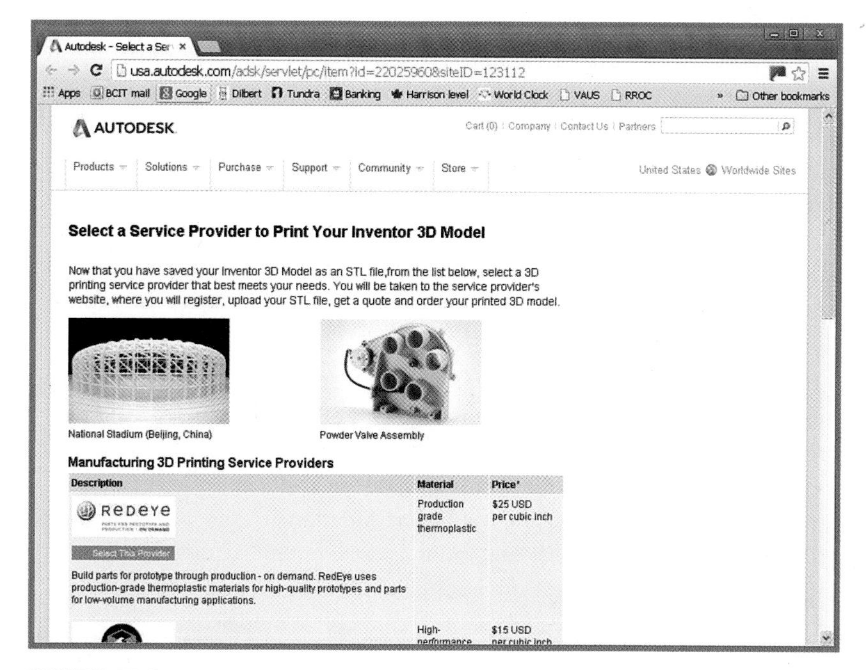

FIGURE 15.2 Selecting a 3D printing service provider.

8. Select a provider, noting that there are two categories: one for manufacturing and one for architectural. The basic difference is that manufacturing parts are usually printed full size, while architectural models are usually scaled down considerably.

9. You are now shifted to the selected service provider's website where you can check out their capabilities and material choices and obtain an "instant" quote. If you like everything so far, you then provide payment and delivery information and hey, presto, like magic, a courier delivers your printout in a couple of days.

Options for 3D Printing

Now that we've seen the basic process for connecting Inventor to a 3D printing service provider, let's look at some of the variables. They include the basic export options we saw earlier, along with some of the 3D print service's requirements:

- Step 5 in the preceding section mentioned the Send to 3D Print Service dialog box, where I told you to click OK. The next time through, however, click the **Options** button first. Surprise! This brings up the exact same **STL File Save As Options** dialog box we saw in Chapter 14. Once again, **Medium** resolution is usually a good starting point for most models and most current printers.
- The STL File Save As Options dialog box also has a **Preview** button that shows you the tessellated model. Unfortunately you can't highlight the tessellation edges the way you can if you open an STL file directly.
- Chapter 14 covered many of the issues to be considered when 3D printing an assembly, not the least of which was the option to print an assembly as one model or as separate, individual parts. The hazard to the latter option is that you can get a great many individual files of common items such as stock fasteners. A solution is to preselect only those components that you actually want to print before starting the Send to 3D Print Service procedure. As per standard Windows selection procedure, you can press and hold the Ctrl key to select multiple specific items, and/or the Shift key to select a group. Unfortunately, this doesn't work when using the Export function.
- Be sure to check out the service provider's minimum and maximum requirements as well. For example, they might require a minimum wall thickness of at least 0.02" (0.5mm), which can be an issue with scaled-down architectural models; minimum running clearances for functional assemblies as I discussed in Chapter 13; and delicate geometry such as a relatively large mass on the end of a thin spindle. In some cases they can print the model but can't ensure safe shipping.
- Some 3D print service companies can offer additional materials and processes such as investment or sand casting and CNC machining.
- Inventor often has alternative ways of accomplishing much the same thing, and 3D printing is no exception. For example, in step 6 Inventor created an STL file to send to the 3D printing service. It's no different from an exported file, and it can be sent to a local 3D printer. Conversely, an exported file can be uploaded to a 3D printing service webpage.

More Inventor STL Export Options

Some might argue that I've saved the best for last. In the following section you'll see additional Inventor functionality that makes the generation of STL files a little more complex but adds options and power.

We start by repeating steps 1–3 from the previous section, but at step 4 we'll diverge a bit from the previous procedure to explore Inventor's additional options and power.

1. Open an Inventor part file.
2. Click the **Application** menu (the big yellow *I*) at the top left of the screen.
3. Hover your cursor over or click the **Print** button.
4. This time, however, click **3D Print Preview**, as shown in Figure 15.3, instead of Send to 3D Print Service as you did previously.

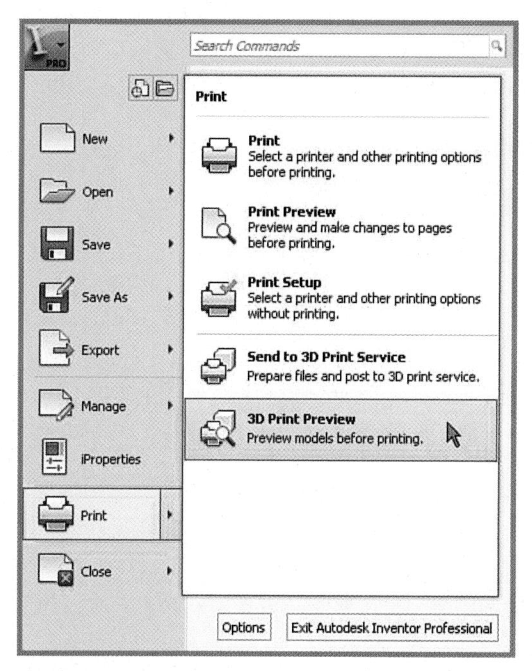

FIGURE 15.3 Starting up Inventor's 3D Print Preview function.

5. This opens a new tab containing a tessellated view of your part, as though you had exported the STL file and then opened it. This tab has a much-abbreviated button bar across the top, as shown in Figure 15.4.

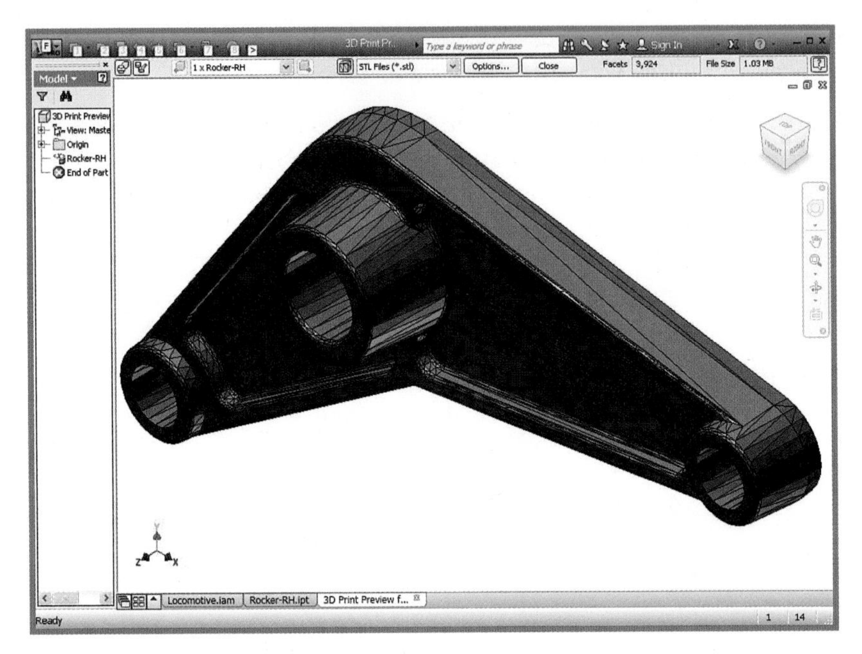

FIGURE 15.4 Activating 3D Print Preview shows the facetted part.

3D Print Preview Button Bar Options

Working from right to left, the button bar contains the following:

- A Help button.
- A window indicating the file size of the current STL file.
- The number of facets that are contained in the STL file.
- A Close button. You can have only one 3D Print Preview tab open at one time, so you have to close the current one before opening another or before returning to a normal part or assembly tab.
- An Options button. Now here's another surprise—it brings up the same STL File Save As Options dialog box we've seen elsewhere. Who would have expected that?
- A file type drop-down list that contains only one entry, which is STL. Perhaps they are leaving room for possible future, as-yet-unknown file types....
- A button to turn facet edge highlighting on and off. Figure 15.4 has facet edges turned on.
- A button, a window, and another button that are applicable only when an assembly is being displayed, and then only under a certain setting. The Help facility refers to these as the Next, Component List, and Previous items, but they aren't actually labelled as such—nor do they have tool tips. I discuss this more later in this chapter.
- A Save Copy As button. This ends up accomplishing the same thing as exporting the file for local printing, as described in Chapter 14.

- Finally, a Send To 3D Print Service button. This shifts you back to the earlier discussion in this chapter, even though you started with the Preview option.

The interesting part about using Print instead of Export comes from the Options button, which launches the familiar STL File Save As Options dialog box. In particular, note what happens when you change the resolution and then click OK. Magic! The screen updates to show the new resolution, the new number of facets, and the new file size.

Figures 15.5, 15.6, and 15.7 show Low, Medium, and High resolutions, respectively. Table 15.1 shows how the facet count and file size increases.

TABLE 15.1 How Resolution Affects Facet Count and File Size

Resolution	Facets	File Size, MB
Low	2,506	0.675
Medium	10,302	2.71
High	24,862	6.54

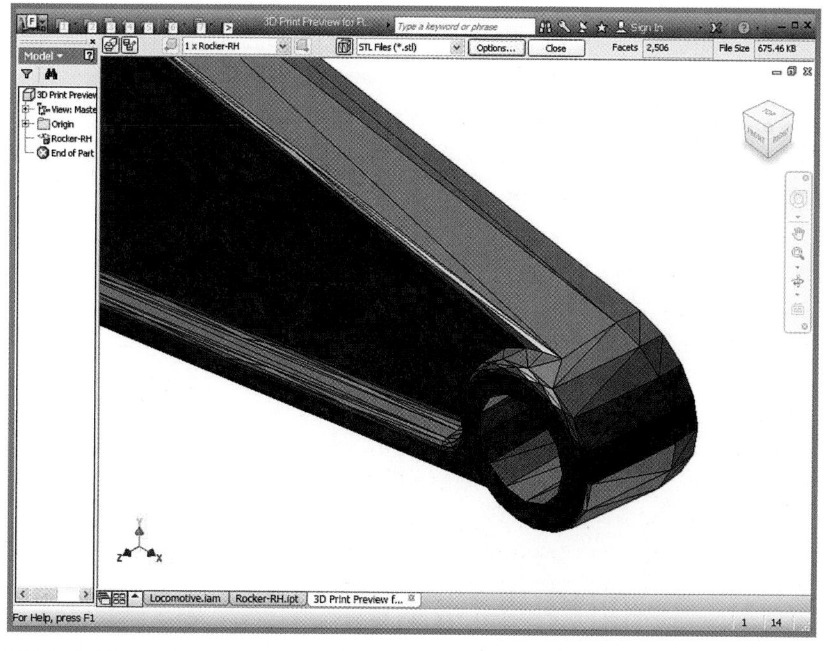

FIGURE 15.5 A low-resolution STL file.

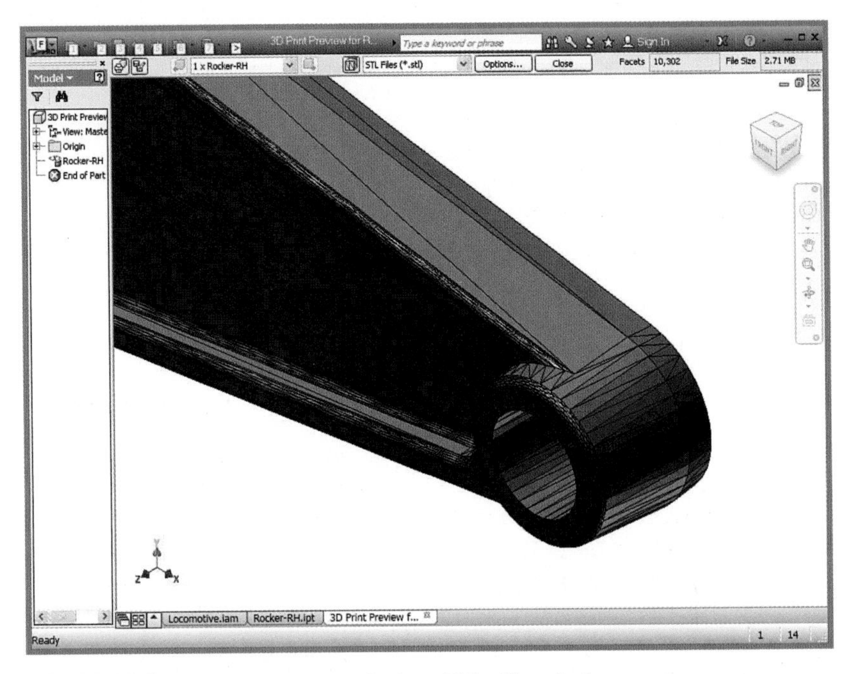

FIGURE 15.6 A medium-resolution STL file of the same part.

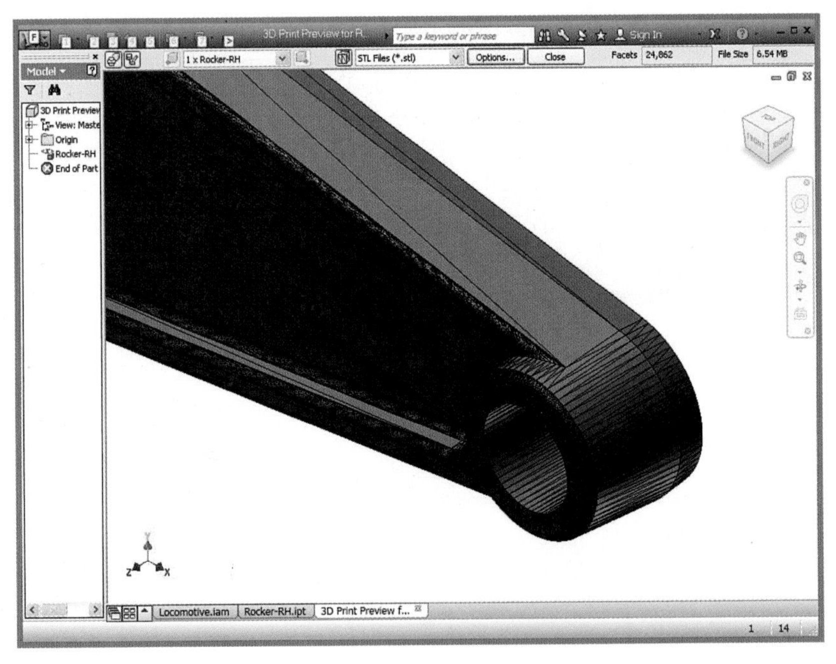

FIGURE 15.7 The same part again at high resolution.

Figures 15.5–15.7 were generated, captured, and saved in about one minute without having to close or open any other files. As you can imagine, this makes it much easier to play with resolutions.

As noted elsewhere in this book, the Medium setting is usually a good starting point unless you are significantly changing the scale factor between the CAD model and the 3D print. The faceting on the 3D print will never be better than what you see onscreen but can be worse depending on printer resolution. In particular, using Custom settings that all approach zero produces government-spending file sizes and geologic processing times.

TIP

Keep in mind that the STL File Save As Options dialog box remembers the last settings used, which means Medium might not be the default. In addition, if you have been playing with super-high-resolution Custom settings, be sure to reset it to Medium before closing the current STL File Save As Options dialog box and then opening even a smallish assembly file. If you don't change these settings, you could spend several hours waiting for your computer to crash because the virtual memory requirements finally exceeded hard drive capacity.

Using the Assembly One File Option

Finally, we come to an area where the Print option works better than the Export option.

If you have the STL File Save As Options dialog box open, close it. Open an assembly file and launch **Print 3D Print Preview**.

Click **Options** to launch the **STL File Save As Options** dialog box. You have an assembly open, so now the **Structure** window drop-down list becomes active:

- **One File**—Displays and exports the assembly as one single item, as shown in Figure 15.8.

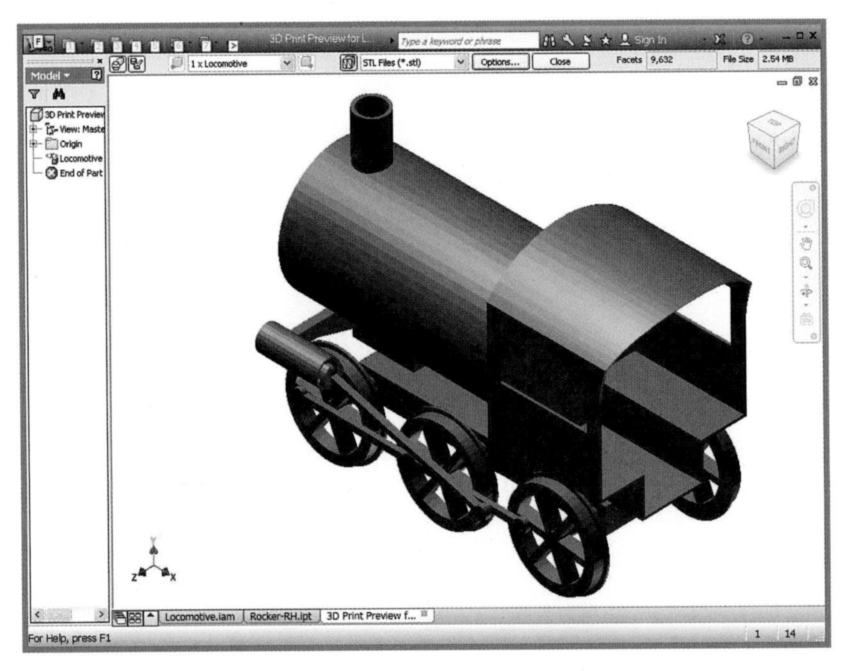

FIGURE 15.8 An assembly in One File mode.

- **One File per Part Instance**—Displays only the top part in the assembly browser structure and exports the assembly as one file per instance of each component. The (unlabelled) component list displays its name and the number of instances of it that occur in the assembly. The buttons on either side of the component list window step you back and forth through the assembly, showing each part in turn and the number of instances of it, while the display screen, number of facets, and file size update accordingly. The component list window is also a drop-down list that displays all components, as shown in Figure 15.9. Click one to jump straight to it.

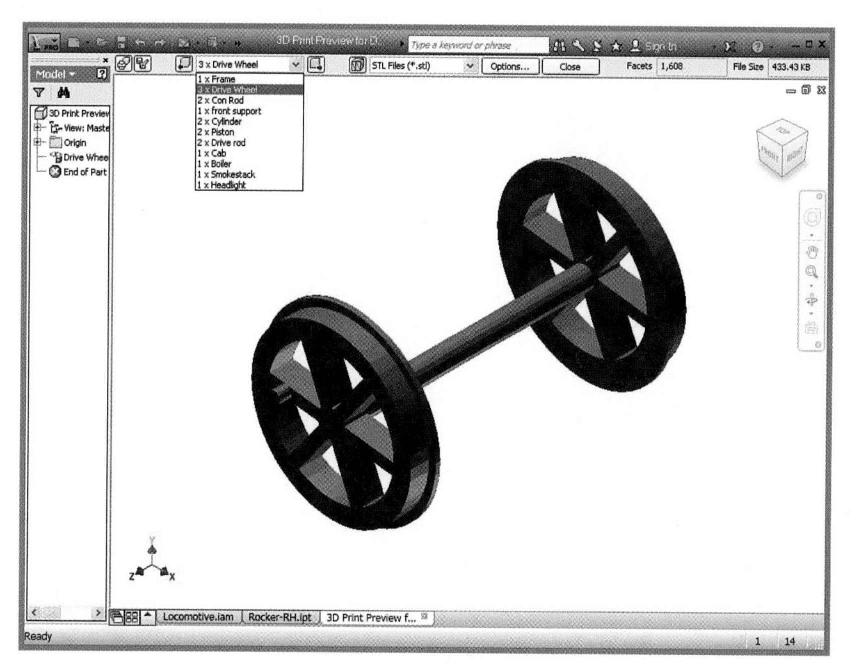

FIGURE 15.9 The same assembly in One File Per Part Instance mode.

NOTE

The Send to Print Service and Save Copy As buttons still export every instance file even if you are looking at only a specific one. The really good news is that it creates a prefix for each filename in the form of **<assembly name>-STL**, thus overcoming the problem I've previously warned about regarding subsequently opening an STL file whose name is the same as its parent part. I wonder why it doesn't do this for single parts?

Summary

As we have seen, Inventor 2014 is amazingly consistent in some areas but a little inconsistent in others. Specifically, what if you want to send only specific components from an assembly to a 3D print service? When using the Export function, you can select specific components before launching Export and only the selected components will be processed. Print, however, doesn't allow this. Instead, you need to go back to the assembly browser, right-click an item you don't want to print, and then click Enabled to turn it off. To speed this up, you can use the Shift and Ctrl keys to select multiple items at one time. You can also use Inventor's Shrinkwrap and Derived functionalities to create simplified versions of your assemblies.

In conclusion, Inventor's 3D Print Preview functionality is usually the easiest and most versatile of Inventor's functionalities for getting a 3D CAD model to a 3D printer—whether it be local or at a service provider.

16

Using a Third-Party 3D Printing Service Bureau

As 3D printing gains popularity in the mainstream, a number of companies have popped up to offer printing services not unlike a service bureau that you'd use for paper copies of documents. These companies typically have a user-friendly online experience and make it easy to get a 3D design printed in a variety of materials. You simply send them your file and then you receive it 3D printed in the material of your choice a few days later.

In this chapter we discuss a number of reasons you might want to use a third-party service for your 3D printing needs, even if you have your own 3D printer. We also discuss how you go about using a third-party 3D printing service.

Reasons to Use a Third-Party Service Bureau

The first (and most obvious) reason for using a third-party service bureau is that you don't own or have access to a 3D printer. You can send your design files to a third-party service bureau from directly within many applications we've discussed in this book.

Using a service is also a great way to print your design in a material not yet available to desktop printers. You can also print your design at much larger scale than is currently offered with desktop printers.

Another reason to consider a service bureau is that many of them have marketplaces in which you can showcase your designs and offer them to the public. This can include selling or giving away your design files as well as selling printed versions of your objects for which the service handles the fulfillment (for a fee).

NOTE

This technology is changing rapidly so not everyone will want to spend their hard-earned money for a desktop machine that might become obsolete before it's paid for. Just like you'd go to Kinko's to get some 2D printing done on paper, a 3D service bureau is perfect for one-off printing jobs where you don't need or want to own the machine.

Capability of Service Bureau Machines

The service bureau machines are generally capable of extremely high-quality printing, using materials and methods not yet available to desktop printers due to high cost, patents, or safety reasons. Many of the more advanced printing techniques, such as laser sintering, are still quite industrial in their processes. Because of this, they require safety precautions, such as ventilation and hazardous waste disposal, and large workspaces. Thus, they are not practical to have in the home or office.

These industrial techniques offer other unique properties when printing your designs, such as the ability to print complex and intricate designs without the need for support material.

Materials that can be used by high-end machines include various acrylics, ceramics, nylons, and even metals. Some of these materials can be printed in a variety of colors and even in full color. The methods used to print your object can vary greatly depending on the service bureau, as can the price.

Players in the Third-Party Service Arena

Now that we've established why you would want to use a service bureau, who are some of the players in this arena? The leaders in this space include i.materialize, sculpteo, and Shapeways.

Autodesk has partnered with a number of these providers and included the ability to order from directly within certain applications. You're not limited to using the partners provided, though. All applications allow you to download your designs, which means you can then upload them to one of the service bureaus.

NOTE

For the purposes of this chapter, I'll walk through my experience using Shapeways to order a 3D print of the cactus I scanned using 123D Capture, in Chapter 4, "Creating 3D Objects with Cameras and 123D Catch."

Uploading 3D Projects to Third-Party Services

Just like desktop 3D printers, the third-party services require a 3D model file, which is typically an STL-formatted 3D model. Most services allow you to send your model files via a simple upload form on their websites. Many of the Autodesk applications covered in this book also enable you to send the model file directly from the application itself.

Ordering a 3D Print

In the menu of 123D Capture (and located similarly in the other 123D applications), select the **Place an Order** option to order a 3D print (see Figure 16.1).

FIGURE 16.1 Ordering a 3D print.

This starts a process of checking the model and takes you to the sizing tab (see Figure 16.2).

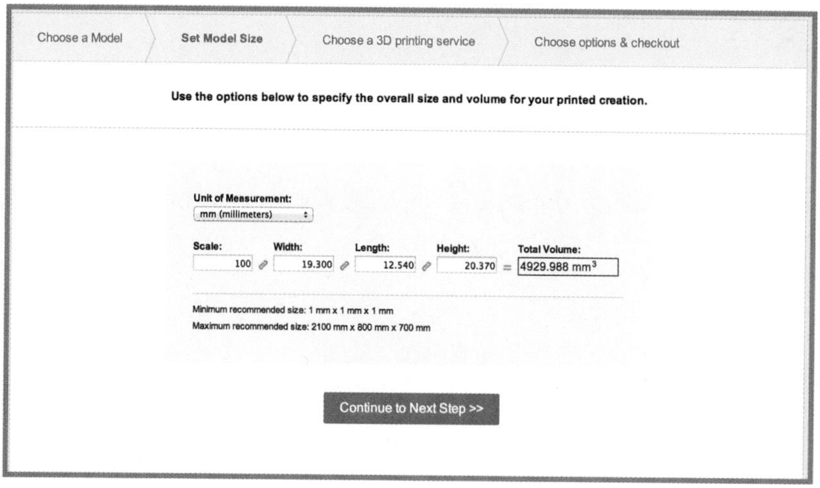

Choose a Model | Set Model Size | Choose a 3D printing service | Choose options & checkout

Use the options below to specify the overall size and volume for your printed creation.

Unit of Measurement: mm (millimeters)

Scale: 100 Width: 19.300 Length: 12.540 Height: 20.370 = Total Volume: 4929.988 mm³

Minimum recommended size: 1 mm x 1 mm x 1 mm
Maximum recommended size: 2100 mm x 800 mm x 700 mm

Continue to Next Step >>

FIGURE 16.2 Sizing your model for 3D printing.

NOTE

This step is important because it sets the size of your model before being sent to Shapeways (or one of the other providers shown in Figure 16.3). You won't have the ability to resize the model after it is sent unless you resend it to the service.

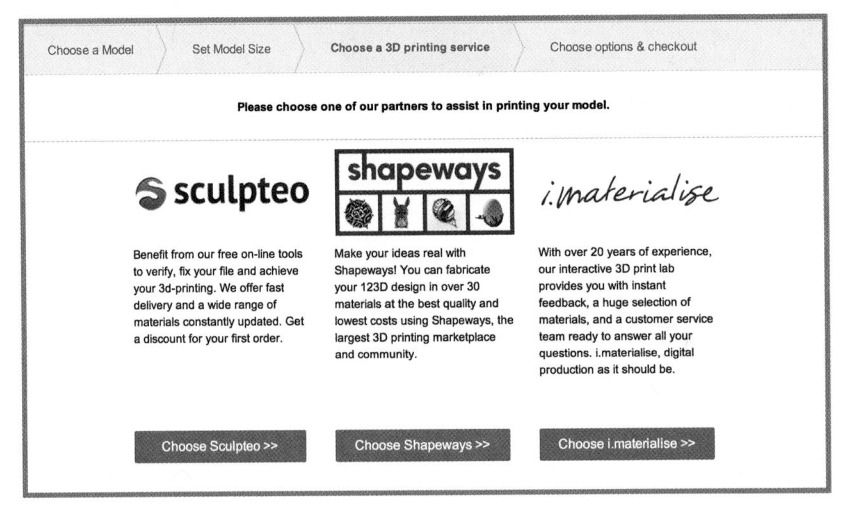

FIGURE 16.3 Choosing a 3D printing provider.

After sizing your model, select a provider to fulfill your printing. For this example, we'll be using Shapeways (see Figure 16.4).

Choose a Model | Set Model Size | Choose a 3D printing service | **Choose options & checkout**

Complete your order through the Shapeways site below*, or go back to choose a different service

shapeways

Your order will be fulfilled by the awesome folks at Shapeways. In order to process your order and keep you informed about its status, we need to know who you are.

Please create a Shapeways account or log in with your existing one.

Log in with My Existing Account

Sign Up

Already have an account? Log in!

Username:

Password:

Confirm Password:

E-mail Address:

I agree with the terms in the Terms & Conditions

I agree with the terms in the Privacy Statement

Create Account

FIGURE 16.4 Sign up for or log in to Shapeways.

After choosing Shapeways, you can sign up for an account or log in if you already have one. This process is similar for the other providers.

After account creation or log in, the 123D browser window then launches a new window or tab and hands over the process to Shapeways (see Figure 16.5).

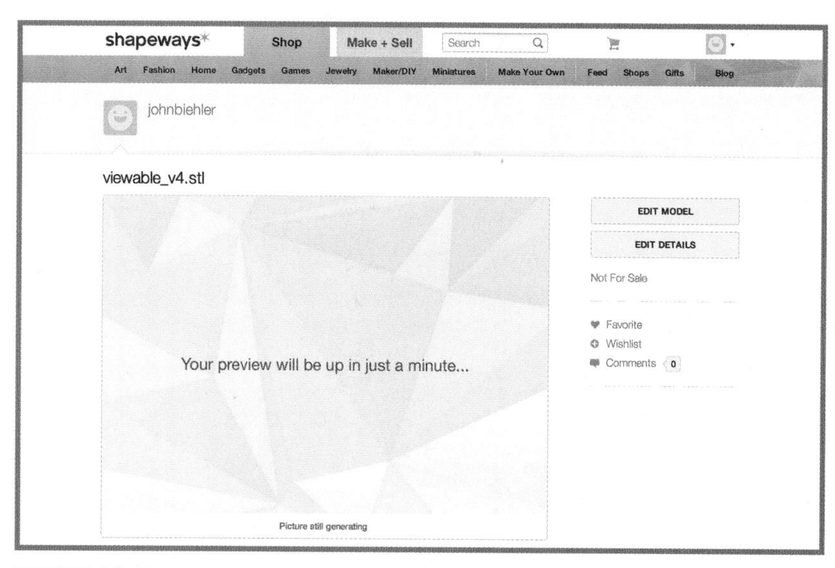

FIGURE 16.5 Loading your model and generating a preview.

NOTE

Depending on the size and complexity of your model, this process can take a few minutes while Shapeways loads your model into its systems.

When Shapeways has finished loading and generating a preview of your model (see Figure 16.6), you'll be able to see the various materials in which the model can be printed and the associated costs.

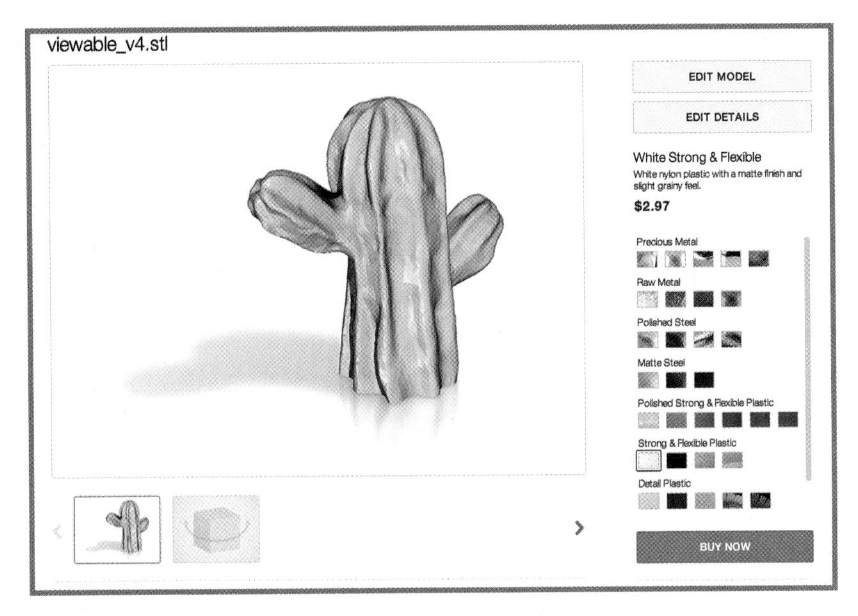

FIGURE 16.6 Previewing the model and materials.

Along with a static preview of your model, you can also view a 3D rotating version of your model to ensure that the importation process hasn't changed it (see Figure 16.7).

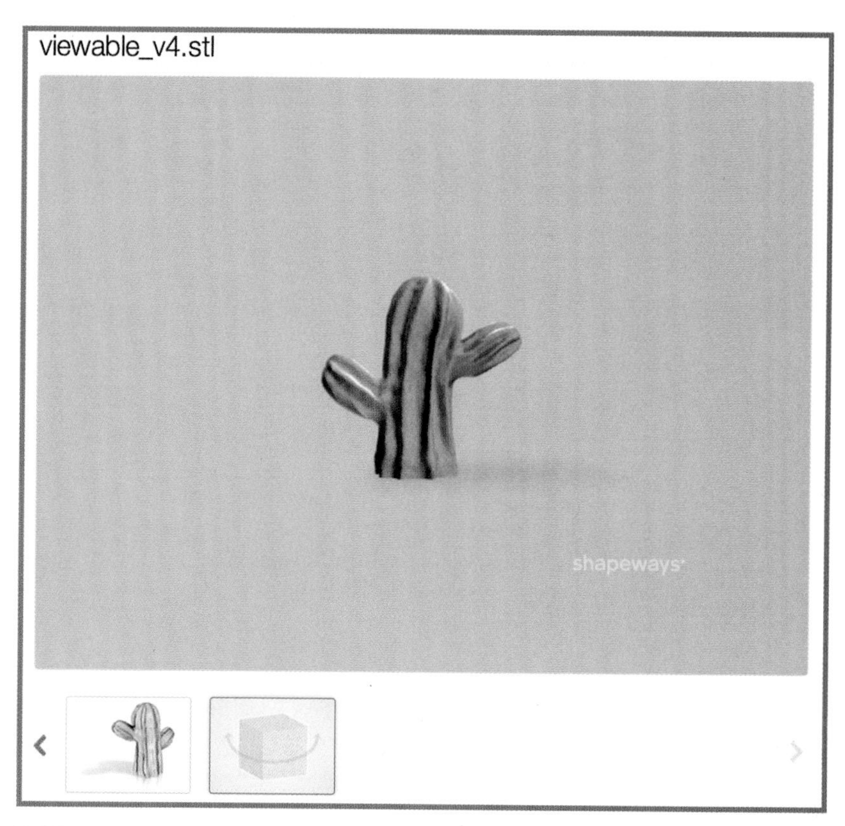

FIGURE 16.7 3D view of your model.

Using the Edit Details Tab

At this point, you should see an Edit Details tab at the top of the screen that will allow you to give your model a name and add a description. You can also categorize your model and add tags if you're planning on making the model public and/or for sale to help prospective buyers find it (see Figures 16.8 and 16.9).

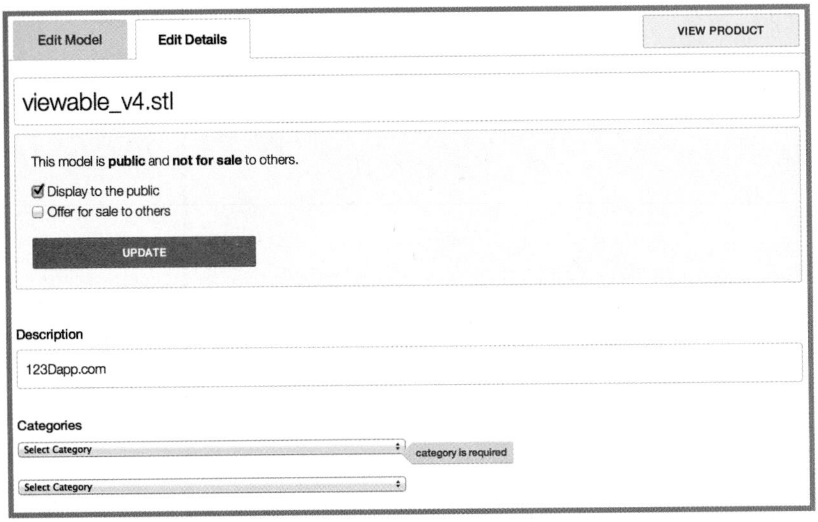

Edit Model
Edit Details
VIEW PRODUCT
viewable_v4.stl
This model is public and not for sale to others.
Display to the public
Offer for sale to others
UPDATE
Description
123Dapp.com
Categories
Select Category
category is required
Select Category

FIGURE 16.8 Setting the details of the model.

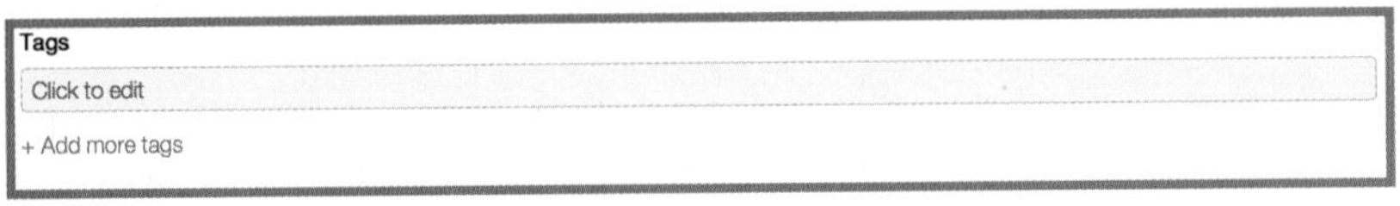

Tags
Click to edit
+ Add more tags

FIGURE 16.9 Adding details and descriptions to your model.

To help showcase your model, you can also upload photos and videos of your model and add captions to the photos (see Figures 16.10 and 16.11).

Photo & Video
Add a photo
Make your products look good on Shapeways!
All images should be sized to 674px wide by 501px tall.
Maximum file size is 2mb.
Choose File
No file chosen
Caption:

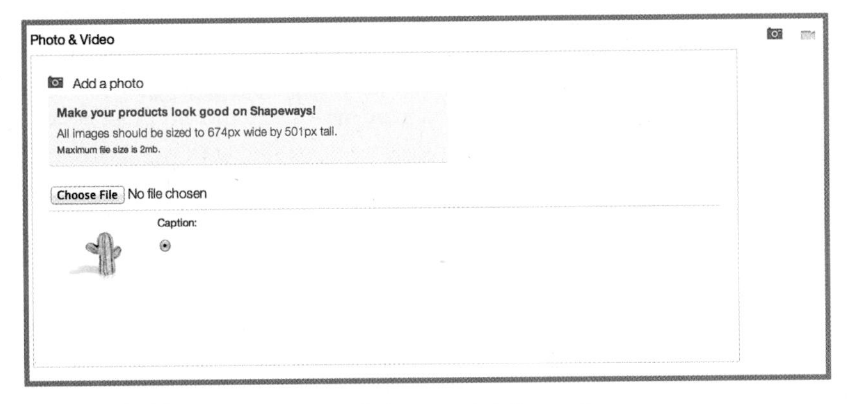

FIGURE 16.10 Add photos of the model if you have any.

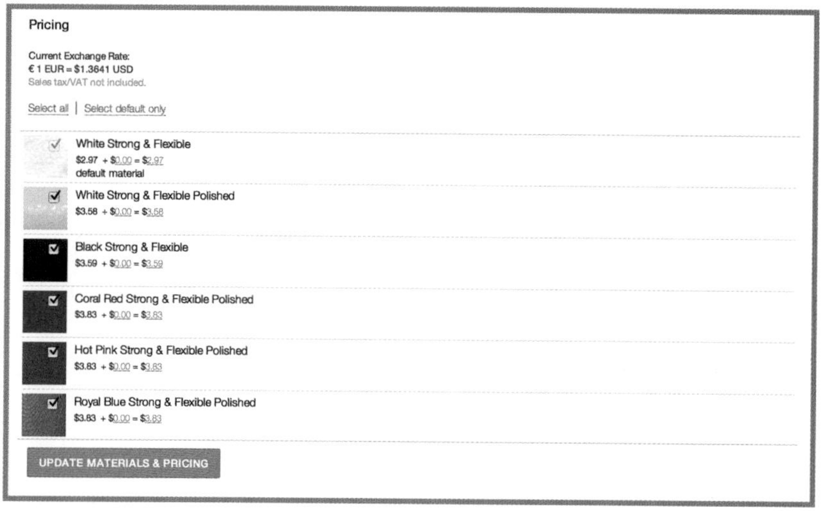

Pricing

Current Exchange Rate:
€ 1 EUR = $1.3641 USD
Sales tax/VAT not included.

Select all | Select default only

White Strong & Flexible
$2.97 + $0.00 = $2.97
default material

White Strong & Flexible Polished
$3.58 + $0.00 = $3.58

Black Strong & Flexible
$3.59 + $0.00 = $3.59

Coral Red Strong & Flexible Polished
$3.83 + $0.00 = $3.83

Hot Pink Strong & Flexible Polished
$3.83 + $0.00 = $3.83

Royal Blue Strong & Flexible Polished
$3.83 + $0.00 = $3.83

UPDATE MATERIALS & PRICING

FIGURE 16.11 Setting the pricing for your model.

Pricing and Materials Options

This section is where you can set the pricing for your model if you've chosen to offer it for sale. This is also where you can see the base cost for printing your model in various materials. As shown in Figure 16.12, your model can be printed with several different materials.

NOTE

If you choose to sell your design, you can add a profit markup to the base rate charged by Shapeways. For example, to print this model in Royal Blue Strong & Flexible Polished (which is a colored nylon material), it costs $3.83 in the size that you uploaded, plus shipping. You can add another amount, which is what you'll get if someone purchases a print of your model from the Shapeways website.

When the model was uploaded, Shapeways also evaluated its printability; as you scroll through the materials list, you can see notes referring to the printability of the model. There are a number of reasons a model isn't printable.

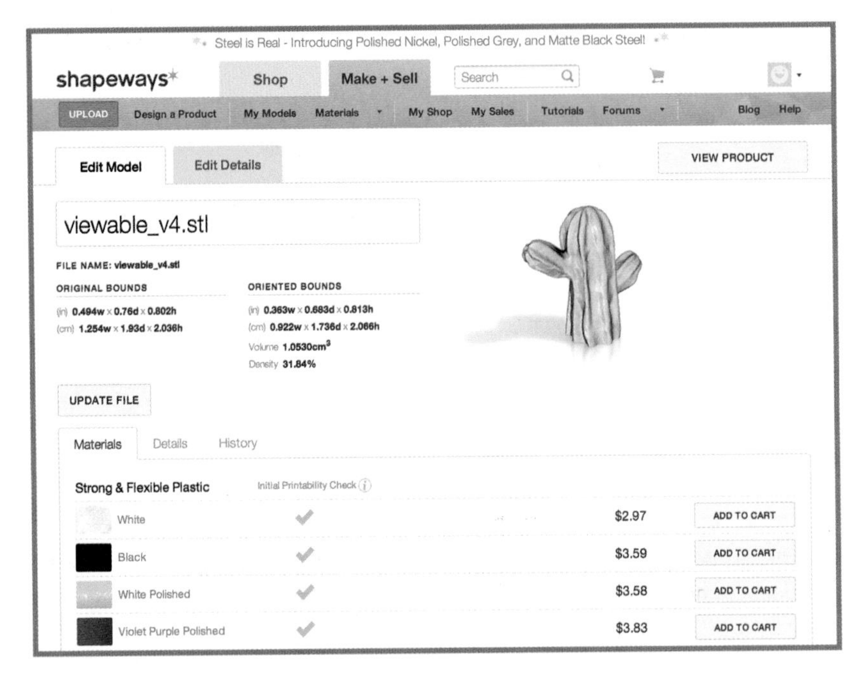

FIGURE 16.12 Previewing the pricing and material options.

In the case of the model used in Figure 16.13, certain materials can be printed only at a certain size. Also, this model is too small for ceramic printing.

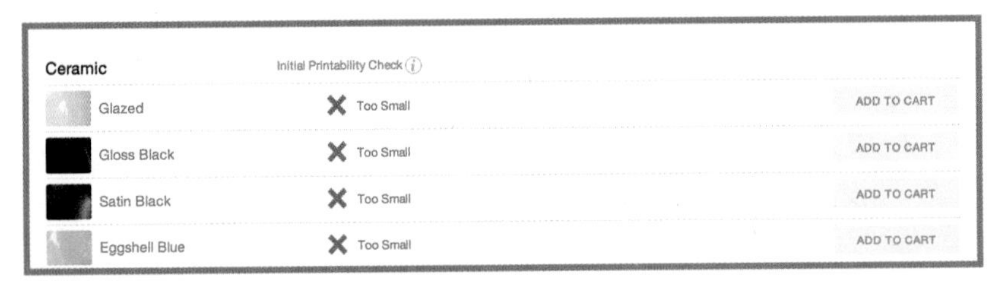

FIGURE 16.13 Pricing and material options continued.

If your model has no color, as in Figure 16.14, printing it in Full Color Sandstone isn't possible (you can still choose regular Sandstone, though).

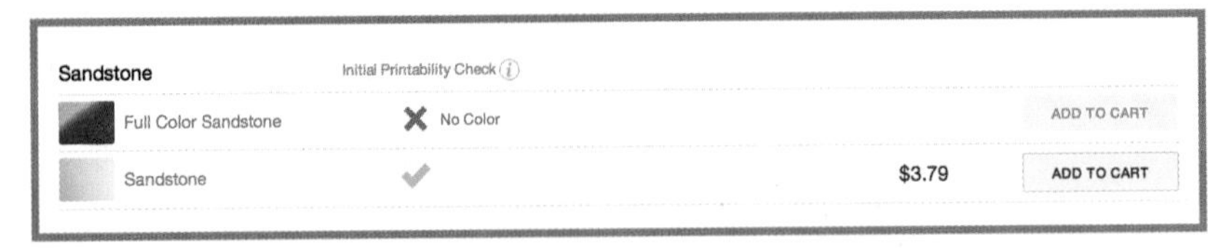

FIGURE 16.14 Pricing and material options continued.

You can see more details about the printability of your model by clicking the **Details** tab located in the **Edit Model** tab (see Figure 16.15).

FIGURE 16.15 Dimensional details about your model.

Clicking the **View Product** tab on the top right takes you back to the model preview screen. From here, you can choose the material in which you want to order the print, add it to your shopping cart, and then proceed to the checkout (see Figure 16.16).

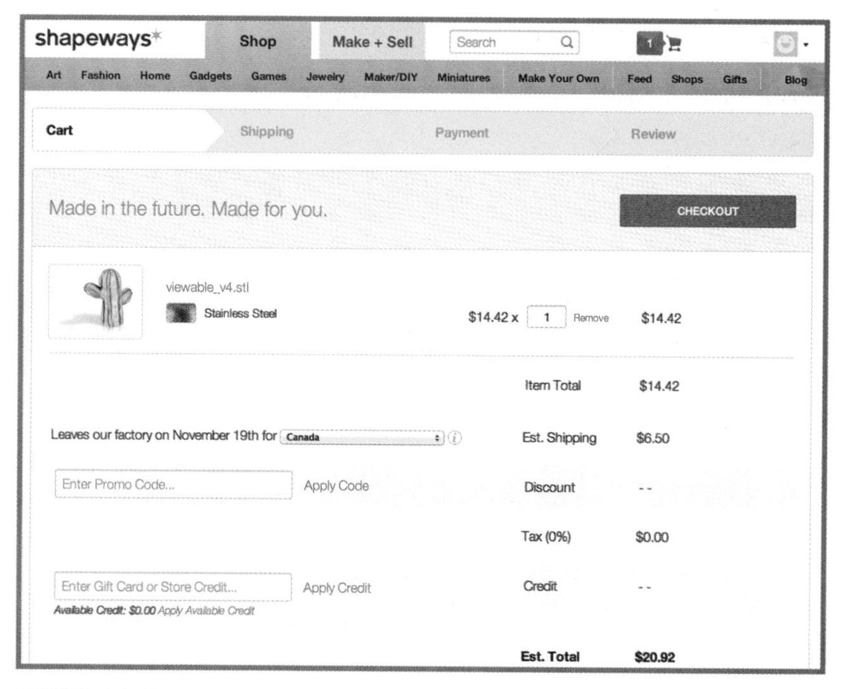

FIGURE 16.16 Shopping cart preview.

Checkout Options

When you go to the Checkout section, you are presented with an estimated delivery time, shipping costs, and the final total for your 3D print.

Click the **Checkout** button, provide a payment option, and your print will start being made.

In most cases, you can get expedited printing and shipping for an additional fee if you can't wait. A few weeks later, the scanned cactus model arrived from Shapeways (see Figure 16.17). It was 3D printed in stainless steel and it turned out perfect!

FIGURE 16.17 Stainless steel 3D-printed cactus from Shapeways.

Summary

In this chapter, we covered a number of reasons you might want to use a third-party service for 3D printing. We also walked through the process of preparing and sending your model to the service, setting up the model for printing, and offering it for sale if you choose.

In the next (and final) chapter, we share our thoughts about the current state of 3D printing and our predictions on what the future holds.

17

The Future of 3D Printing

Predicting the future is notoriously inaccurate, especially when it comes to technology. Where are the flying cars that 1950s *Popular Science* magazine promised us "in the near future"? And vacations at space resorts on the moon? On the other hand, who predicted that fax, audiocassettes, VHS, CDs, and DVDs would disappear so quickly?

On the other, other hand "picture phones" that let you see the other person were still science fiction 10–15 years ago, but now we regularly Skype our son and his wife on another continent from our home, or from wherever we can get an Internet connection. And it's free!

The Future According to Bill

Having said all that, here are my predictions for the future of 3D printing.

Most 3D printers produce parts made of a small range of plastics and corn starch, with a typical build volume of a cubic foot or smaller. I predict that the range of materials and the size of printers will change significantly to include the following:

- Hardened steel for things like custom chain sprockets for racing bicycles
- Aluminum engine blocks for experimental and racing cars
- Full-size sandstone decorative architectural panels
- Full-size concrete houses

I also predict that the following areas will change dramatically because of 3D printing over the next decade:

- **Food**—Long-term space missions such as going to Mars will carry an inventory of basic food ingredients that the astronauts will be able to print to order different meal items such as pizza, bread, doughnuts, steaks, hamburgers, and so on.
- **Human organs**—Doctors will extract human stem cells from a patient's bone marrow, reproduce them in the laboratory, 3D print a replacement organ, and transplant it back into the patient. There will be no need to use anti-rejection drugs because the organ will be made up of the patient's own cells.
- **Medicines**—A stock of basic chemical compounds will be used to 3D print custom, build-to-order medications.

Professionals and home hobbyists alike will see changes, including these:

- Printer prices will drop down to the $300 range and lower.
- 3D printers will reproduce, printing many of the parts needed to build another 3D printer.
- 3D printers will be able to produce full-size display models of things like aircraft jet engines and motorcycles.
- You'll be able to take several photographs of an object and then 3D print them into a physical copy of the original object.

The noted science-fiction writer William Gibson (*Neuromancer* and *Johnny Mnemonic*) has said, "The future is already here, it's just not evenly distributed." Everything listed so far already exists or is being actively developed, and many of them have already been mentioned in this book. Rule #1: shoot first, and then declare what you hit to have been the target.

The final item on my list—liability lawyers will go into a feeding frenzy. Electronic recording and playback, including Internet distribution, shook up existing copyright laws. Electronic circuits and computer programs have already spawned significant changes to the patent industry.

Well, I believe that you ain't seen nothin' yet. For example, if someone downloads a file(s) from the Internet, 3D prints an object, and then someone is injured or killed by it, who is liable? The original designer? The company whose software was used to design the object and/or to drive the printer? The webpage that hosted the files? The 3D printer manufacturer? The supplier of the raw material for the printer? The person who did the actual 3D printing? The person who was using the object? Perhaps some sort of design certification agency needs to evolve.

There is a phrase that has been bandied about in the CAD and 3D printing industries to the effect that they want to "democratize the design process." You will no longer be tied to the tyranny of what some designer or manufacturer thinks you might buy; instead you'll be able to design what YOU want.

The problem I see with this is that "design" involves an awful lot more than pretty shapes. Consider something as mundane as a door knob. Can't find a style you like? No problem. Simply design and print what you want. Okay, but have you considered the appropriate government and industry standards? How about fire ratings? Do you have an American Society for Testing and Materials (ASTM) standard test door in your basement so you can test your design for strength and durability? I spent 27 years designing doorknobs, and I'm a graduate mechanical engineer.

Having said all that, I believe that things will shake out and that the matter replicator is here to stay in one form or another. The Wright brothers flew and men landed on the moon within the lifespan of my oldest uncle. By comparison, the 3D printing industry is still at the level of Lilienthal's hang gliders.

Sorry, there are no pretty pictures for this section. I tried setting the date and time in my camera to 10 years ahead, but that didn't work. I even tried the International Date Line trick of having someone in Australia take a picture one day ahead; still no joy.

The Future According to John

The current state of 3D printing is hard enough to put a finger on. With its ever-changing and evolving status, it can also be hard to nail down what the future will hold. I've been heavily involved in 3D printing since 2011 when I built my first 3D printer from laser-cut wood and hand-soldered electronics. In just these few short years, things have changed dramatically.

There are, however, a few things that can be expected to happen as more people and companies embrace the technology:

- Every day a new crowd-funded campaign for a different 3D printer design appears with a fresh take on the technology, which is constantly pushing the envelope.
- Lower-cost machines and raw materials will be available thanks to the economies of scale.
- Software will improve and be easier to use (as we've tried to demonstrate with this book).
- Innovative printing materials will allow for more practical applications of 3D-printed objects.
- All these elements will help propel the technology forward and into the minds of the general public.
- The explosion of interest in 3D printing technology in the last few years has been primarily due to the open source community and its involvement in spreading the technology far and wide, into corners that might not have normally adopted this kind of technology.

  When you bring new people into a room with a new technology, all kinds of great things can happen, especially when they don't have a background in that specific technology (sometimes referred to as "baggage"). The open source community also comprises some of the most passionate users of the technology who iterate and innovate literally daily in the space.
- As was mentioned earlier in this book, 3D printing is actually composed of a number of technologies. The most commonly described being fused deposition modeling (FDM) because it's the least expensive and currently most accessible technology available.
- With the patents soon expiring (likely already by the time you read this) on selective laser sintering (SLA), it seems inevitable that the next wave of low(er)-cost 3D printers will be SLA based and likely replicate the pricing and availability of FDM printers today, at a higher resolution and quality.
- SLA technology also makes it much easier to bring different materials into the mix. There are already a number of projects working on lowering the cost of entry for printing with

various metals and other previously expensive raw materials. These powders and resins can result in printed objects that are extremely strong and durable; more natural materials like flexible rubber and silicone, which currently can only be used in injection molding applications, will be possible.

Time will tell if 3D printing can happen inexpensively enough while being safe to work with in a nonindustrial environment like the home or office. This will happen at a very rapid pace thanks to the foundational work and media exposure, again in large part due to the open source community's work with FDM.

3D printing is already in nearly every possible industry, and its use will continue to grow as the costs go down and more people are exposed to and get their hands on the technology.

Indeed, it's a very exciting time to be involved in 3D printing.

Index

Symbols

A

B

C

D

E

F

G

H

I

J

L

M

N

O

P

Q

R

S

T

U

V

W-X-Y-Z